D0759791

THE XERCES SOCIETY GUIDE

Farming with NATIVE BENEFICIAL INSECTS

Ecological Pest Control Solutions

Eric Lee-Mäder
Assistant Director, Pollinator Program, The Xerces Society

Jennifer Hopwood
Pollinator Conservation Specialist, The Xerces Society

Lora Morandin
Morandin Ecological Consulting

Mace Vaughan
Pollinator Program Director, The Xerces Society

Scott Hoffman Black
Executive Director, The Xerces Society

The mission of Storey Publishing is to serve our customers by publishing practical information that encourages personal independence in harmony with the environment.

Edited by Deborah Burns
Art direction and book design by Cynthia N. McFarland
Text production by Liseann Karandisecky

Cover photographs by © Don Keirstead, NH NRCS (front, right), © Jessa Cruz, The Xerces Society (back), © Richard Greene (green lacewing, front, top left), and USDA-NRCS/Lynn Betts (front, bottom left).
Interior photography credits appear on page 257
Illustrations by © Marjorie C. Leggitt
Graphs by Ilona Sherratt

Indexed by Samantha Miller

Storey Publishing
210 MASS MoCA Way
North Adams, MA 01247
www.storey.com

Printed in China by Toppan Leefung Printing Ltd
10 9 8 7 6 5 4 3 2 1

Library of Congress Cataloging-in-Publication Data
Farming with native beneficial insects : ecological pest control solutions / by the Xerces Society.
pages cm
Other title: Ecological pest control solutions
Includes bibliographical references and index.
ISBN 978-1-61212-283-0 (pbk. : alk. paper)
ISBN 978-1-61212-284-7 (ebook) 1. Insects as biological pest control agents—United States. 2. Beneficial insects—United States. I. Xerces Society. II. Title: Ecological pest control solutions.
SB976.I56F37 2014
632'.7—dc23

2014012568

For our kids — past, present, and future

◆ ◆ ◆

Countless wonderful folks contributed to the creation
of this book, including some of the most forward-thinking farmers,
conservationists, scientists, and insect photographers of our time.

We thank all of them, and feel especially compelled to call out
the following supporters and contributors:
Matthew Shepherd, Margo Conner, Ashley Minnerath,
Claire Kremen, Sarina Jepsen, Jessa Guisse Cruz, David Biddinger,
Whitney Cranshaw, Rachael Long, Tessa Grasswitz, Debbie Roos,
Glynn Tillman, Vilicus Farm, Grinnell Heritage Farm,
The Kerr Center, John Tooker, John Anderson, Rufus Isaacs,
Don Keirstead, Brett Blaauw, Matt O'Neal, Gwendolyn Ellen,
Jim Lerew, Ed Rajotte, Chris Helzer, David James,
Nancy Lee Adamson, Thomas Ward, and Alex Wild

CONTENTS

Preface x

PART 1: Beneficial Insect Ecology 1

1. Pest Control with Beneficial Insects 2

What Are Beneficial Insects? • Pest Control and Farming • Enhancing Beneficial Insect Populations • Approaches to Biocontrol • Common Predators and Parasitoids: Their Habitat and Prey

2. Why Farm with Native Beneficial Insects? 22

Pest Control • Benefits beyond Pest Control • Case Study: Pest Management in Washington State Vineyards • Case Study: Milkweed, Stink Bugs, and Georgia Cotton

3. Evaluating Beneficial Insect Habitat 30

Habitat Essentials • Farm Practices Checklist • Case Study: Beneficial Insects Save Christmas

PART 2: Improving Beneficial Insect Habitat 43

4. Designing New Beneficial Insect Habitat 44

What Beneficial Insects Need • Habitat Size and Location • Wildflower Selection

5. Native Plant Field Borders 56

Establishing Borders from Seed • Site Preparation • Seeding • Long-Term Maintenance • Sample Seed Mixes for Native Plant Field Borders

6. Insectary Strips 82

Perennials or Annuals? • Plant Insectary Strips • Case Study: An Insectary Seed Mix for New Mexico Pumpkins • Sample Insectary Seed Mixes

7. Hedgerows 92

Installing a New Hedgerow • Revitalizing Old Fencerows • Sample Hedgerow Plant Mixes • Case Study: Hedgerows on California Central Valley Farms

8. Cover Crops 115

Species Selection • Establishing a Cover Crop • Case Study: A Better Farm for Beneficials

9. Conservation Buffers 130

Contour Buffer Strips • Grassed Waterways • Riparian Buffers and Filter Strips • Shelterbelts and Windbreaks • Organic Farm Buffers • Case Study: Buffer Strips for Soybean Aphid Control

10. Beetle Banks and Other Shelters 143

Beetle Banks • Tunnel Nests • Wasp Shelters • Brush Piles • Insect Hotels • Case Study: Banking on Beetles in Oregon

PART 3: **Managing Beneficial Insect Habitat** 159

11. Reducing Pesticide Impacts 160

If You Must Use Pesticides • Pesticide Selection • Microbial Insecticides and Nematodes • Alternatives to Pesticides • Controlling Spray Drift • Case Study: Designing Windbreaks to Limit Pesticide Drift • Case Study: Biological Mite Control in Pennsylvania Apple Orchards

12. Long-Term Habitat Management 174

Disking, Mowing, and Burning • Grazing • Rotating Habitat Disturbance • Interseeding Wildflowers • Case Study: Beneficial Insect Habitat on an Oklahoma Farm and Ranch

PART 4: **Common Beneficial Insects and Their Kin** 183

PREDATORY INSECTS 184

Assassin Bugs, Ambush Bugs 184

Big-Eyed Bugs 185

Damsel Bugs 186

Minute Pirate Bugs, Insidious Flower Bugs 187

Predatory Stink Bugs 188

Mantids 189

Green Lacewings, Brown Lacewings 190

Checkered Beetles 191

Firefly Beetles, Fireflies, Lightning Bugs 192

Ground Beetles 193

Tiger Beetles 194

Lady Beetles, Lady Bugs, Ladybird Beetles 195

Soft-Winged Flower Beetles 196

Soldier Beetles 197

Rove Beetles 198

Flower Flies, Hover Flies 199

Predatory Wasps 200

PARASITOIDS 201

Parasitoid Wasps 201

Scarab-Hunting Wasps 202

Tachinid Flies 203

NONINSECT BENEFICIAL PREDATORS 204

Jumping Spiders, Wolf Spiders, Orb Weaver Spiders, Sheet-Weaving Spiders 204

Harvestmen 206

Predatory Mites 207

PART 5: Plants for Conservation Biocontrol 209

Native Wildflowers 210

Native Flowering Trees and Shrubs 220

Native Grasses 226

Cover Crops and Nonnative Insectary Plants 230

PART 6: Appendix 235

Additional Resources 236

About the Authors 242

The Xerces Society for Invertebrate Conservation 243

Index 245

Preface

Lacewing larvae contribute to pest management on farms and in gardens by attacking aphids and other pests.

Native insects that prey upon or parasitize crop pests are unsung champions in agricultural systems. Although vast numbers of such beneficial insects are at work on farms across the world, they are overlooked, undervalued, and eclipsed in the public imagination by a comparatively tiny number of pest species. Yet, as the case studies scattered throughout this book reveal, farmers as diverse as Christmas tree growers in Illinois, orchardists in Pennsylvania, and wine grape growers in the Pacific Northwest benefit from the natural pest control provided by beneficial insects.

This book was developed to raise awareness of these helpful animals. Here we discuss their ecology and offer realistic strategies for conserving and enhancing them on working farms. While farming is our main focus, the strategies we describe throughout can be effectively scaled down for the home gardener as well. In this book you'll learn why you should conserve beneficial insects, how they can help control pests, and how you can protect and restore beneficial insect habitat. You'll also find additional information on the insects themselves, specific native

The strategies outlined in this book will help you take unused areas of the farm like this fence line, which had been managed for years with herbicides to control weeds, and convert them into functional beneficial insect habitat, as seen on the opposite page.

plants that support them, and USDA financial resources to help implement conservation measures on your farm.

While native beneficial insects alone may not solve your major pest problems (although in some cases they can!), ongoing research across the country clearly demonstrates a very strong link between conservation of natural habitat and reduced pest problems on farms. Where native beneficial insects are present they can:

- Reduce the need for insecticides
- Improve crop yields by reducing pest damage
- Reduce the need to release nonnative beneficial insects

In addition, the conservation practices that support native beneficial insects can:

- Benefit diverse pollinators, such as native bees and managed honey bees
- Provide habitat for wildlife, such as game and songbirds
- Reduce weedy plants on field edges
- Function as buffer systems that reduce erosion, improve water quality, and make your land visually pleasing
- Help organic farmers meet biodiversity conservation requirements for USDA organic certification

The native wildflowers established along this fence line contribute to pest control by providing habitat for beneficial insects that move out into the adjacent crop field. The flowers also support pollinators, are beautiful to look at, and are not weedy.

The concept of providing habitat for native insects that attack crop pests is referred to as **conservation biological control**, or often simply **conservation biocontrol**. While we describe other biocontrol practices in the first chapter, we focus this book specifically on conservation biocontrol because it represents a win-win conservation opportunity: a chance to reduce pest damage to crops while at the same time supporting native wildlife and biodiversity.

This book provides a broad overview of the principles of conservation biocontrol, as well as an introduction to the important groups of beneficial insects that you may find on your farm. Because pests and beneficial insects vary among cropping systems and regions, it is difficult to provide specific advice that will control pests in all situations. Additionally, we do not always know the thresholds at which some beneficial insects will provide pest control for a given cropping system. Given these limitations, this book does not offer complex guidelines tailored to specific crops or regions. Rather, the general strategies that we outline are applicable to a wide range of farms. We hope that the more specific strategies described in the case studies throughout will inspire you to develop management practices tailored to your own needs.

We've been amazingly lucky to work with a huge range of farmers and ranchers across the country who are actively managing their land for beneficial insects: berry producers in Maine, Massachusetts, and Florida, alfalfa seed producers in Wyoming, ranchers in Oklahoma, nut and vegetable producers in California, and tree fruit producers in the Pacific Northwest and the Great Lakes — as well as nearly every conceivable type of farm in between. Many of the conservation biocontrol strategies described here are ones that we learned from those farmers. That knowledge, combined with the research of many excellent scientists and support offered by the USDA Natural Resources Conservation Service, continues to demonstrate that conservation biocontrol can be an effective and practical approach to pest management, while also contributing to the broader sustainability of farms and ranches.

Conservation biocontrol is no magic bullet. For as long as we grow crops, pests will be part of the equation. But incorporating conservation biocontrol into Integrated Pest

Beneficial insect habitat can be integrated into farms of all sizes, from small urban farms to 1,800-acre dryland grain farms, such as this one in Montana.

Management represents a complementary approach for farmers who want to use fewer chemicals. Whether you are a conventional or organic producer, and whether you want to eliminate pesticides entirely or just save money by spraying less, conservation biocontrol can become part of your farm system.

You may achieve the higher beneficial insect numbers you want simply by protecting noncrop areas on your farm. Creating new beneficial insect habitat can increase those numbers even more, while also providing beautiful new recreation areas for you and your family. Combining those approaches with pesticide alternatives such as physical barriers may allow you to market a premium product, tell your customers that your farm is increasingly sustainable, and serve as a role model to other farmers.

We hope that you'll take this guidance, adapt it, build upon it, and in turn share with us your own experiences. Happy farming!

PART 1

Beneficial Insect Ecology

IT CAN BE EASY TO OVERLOOK beneficial insects. Many are tiny or cryptic, and others are easily mistaken for pests. Despite their fascinating and varied natural histories, they live their lives quietly behind the scenes.

But we must not discount the role that beneficial insects play. Native predator and parasitoid insects provide ecological pest control services estimated to be worth at least $4.5 billion annually in the United States.

To provide biocontrol services, beneficial insects need stable habitat that can supply supplemental food, alternate prey or hosts, and shelter. Here you'll learn of the many ways to offer habitat to support beneficial insects.

Beneficial insects such as lacewings are highly dependent on native plants, especially wildflowers, to thrive. Crop plants alone are typically not enough to sustain their populations and enhance their ability to control pests.

1

Pest Control with Beneficial Insects

- Most types of insects are beneficial to humans.
- Biological control, or biocontrol, focuses on the beneficial insects that attack crop pests.
- The main groups of beneficial insects that provide pest control are predators and parasitoids.
- Conservation of beneficial insects complements Integrated Pest Management.

INSECTS PLAY many valuable roles in our environment. Though we tend to focus our attention on pests, the vast majority of insects are working with us, rather than against us. Insects recycle nutrients, help decompose plant and animal waste, contribute to soil quality by churning and aerating soil particles, provide pollination, attack crop pests, and are food for fish, songbirds, and other wildlife.

While any insect that provides a service to humans could be considered a "beneficial insect," in this book we are referring to predatory and parasitoid insects and a few related animals, such as spiders.

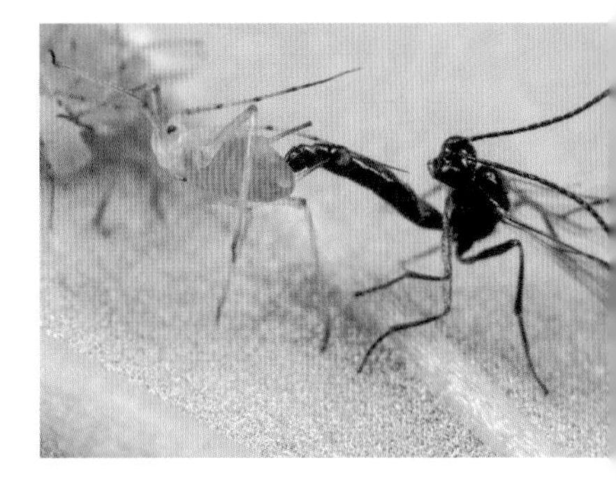

A parasitoid wasp injects her eggs into an unfortunate aphid. The wasp's growing young will slowly consume the still-living aphid host.

What Are Beneficial Insects?

Beneficial insects include predatory insects such as lady beetles (ladybugs), parasitoid insects such as some wasps, pollinators such as bees, and various soil-dwelling insects that contribute to crop health. All of these can provide important services to farmers. This book focuses on two main types of beneficial insects: predators and parasitoids, and their role in natural pest management.

In general, we consider insects beneficial if they perform ecosystem services. In the context of this book, however, we use the term **beneficial insects** to describe those insects that suppress crop pests: insect predators and parasitoids, as well as several noninsect predatory arthropods such as spiders. Others may use the term "natural enemies" to describe these groups. A natural enemy is simply an organism that feeds on another organism. Insects or other arthropods that contribute to pest control may also be referred to as "biocontrol agents." Some people prefer one term over another, but the intention behind all three is similar: these insects and arthropods provide us with a beneficial service that should not be overlooked.

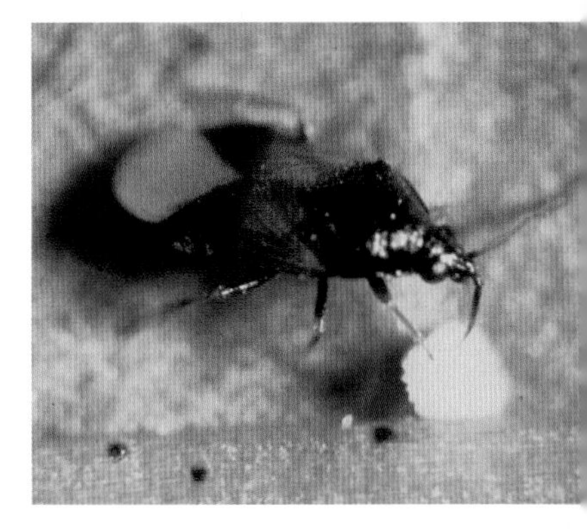

Minute pirate bugs may be tiny in size, but their contributions to pest control are not small.

*This convergent lady beetle (*Hippodamia convergens*) and other lady beetles are beneficial predators as larvae and adults.*

COMPARING METAMORPHOSES

Different insect groups go through different stages of development to reach adulthood.

Some groups of insects go through incomplete metamorphosis, *a gradual, simple transition from egg to nymph to adult. Nymphs frequently resemble smaller versions of adults, though they do not possess wings and cannot yet reproduce.*

Other groups undergo complete metamorphosis, *developing from an egg to a larva (which looks very different and often occupies a different habitat than the adult) to a pupal stage, during which metamorphosis occurs, to an adult.*

The life cycle of a true bug (minute pirate bug, left), which involves incomplete metamorphosis, is here compared with that of a lacewing (below), which undergoes complete metamorphosis.

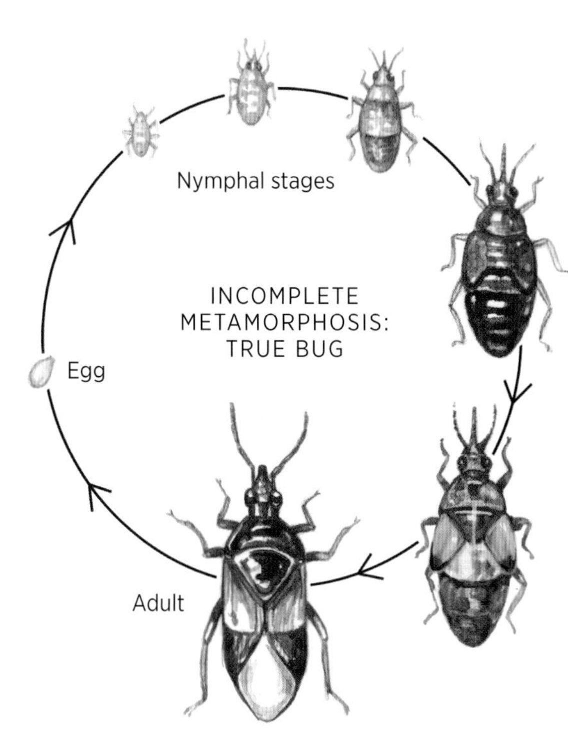

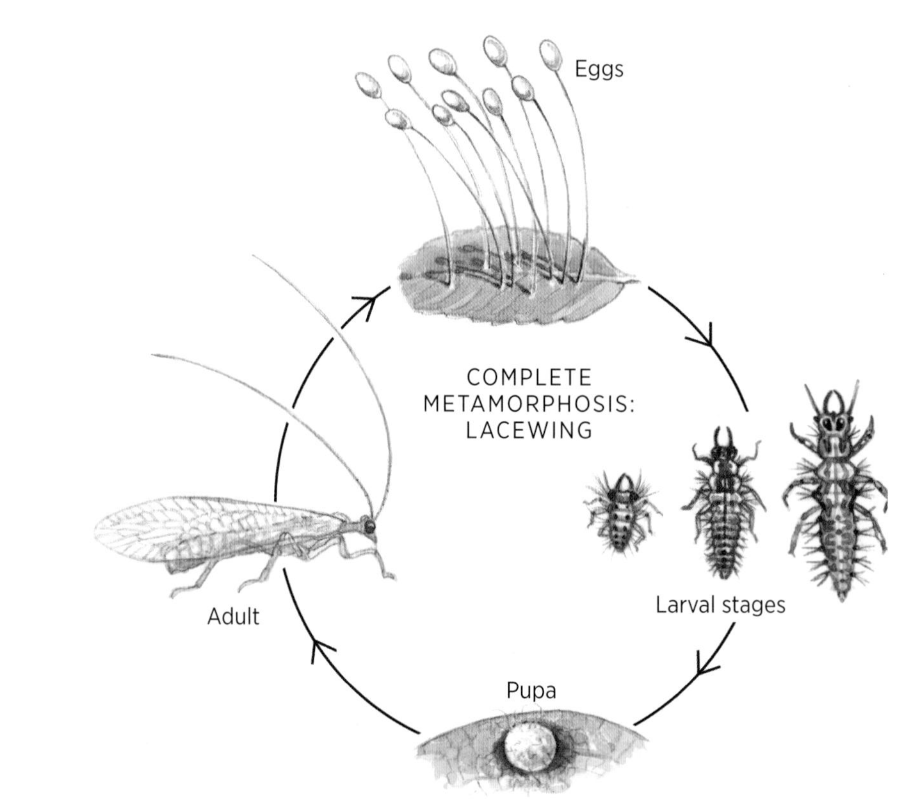

*An orb weaver spider (*Mangora *spp.) waits for an insect to get caught in her intricate spiral web.*

Predators

Predators are animals that attack and consume their prey. Within the context of biocontrol, predators hunt, kill, and consume pests. The most familiar predators of economic importance to farmers are insects, such as lady beetles and wasps, but spiders and predatory mites, nematodes, birds, and bats can also be important predators of crop pests. In this book, we focus on the role of predatory arthropods (insects, spiders, and mites) and ways to enhance their populations.

Characteristics of predatory arthropods include:

- They can range in size from mites the size of the period at the end of this sentence to 4-inch-long mantids (praying mantises).
- Most predatory arthropods are generalists, meaning they feed on a wide array of pests.
- Individual insects can consume many times their weight in prey: lacewing larvae, for example, can eat hundreds of aphids a week.
- Even if they don't seem plentiful in a field, predatory insects can still significantly reduce the number of pests.

Some assassin bugs hide under cover to ambush prey; additionally, the forelegs of Zelus *spp. exude a sticky substance that also helps them to trap prey.*

The diversity of predators in a field crop can be enormous. For example, more than six hundred species of predators in 45

A tiny parasitoid wasp parasitizes a hornworm caterpillar. Soon the eggs inside the caterpillar will hatch and the larvae will begin to feed, undetected by the caterpillar's immune system.

When wasp larvae reach maturity, they emerge from within the still-living caterpillar and pupate within silken cocoons.

Free-living adult parasitic wasps emerge from the cocoons, leaving behind silken cases and a significantly weakened host.

By the time most adult wasps have emerged from their cocoons, the host caterpillar has died or will die soon.

insect families and 23 families of spiders and predatory mites have been recorded in Arkansas cotton fields. In the northeastern United States, scientists have documented 18 species of predatory insects in potato fields. Although the impact of any one type of predator may be minor, their combined effect on pests can be considerable.

Parasitoids

Animals that live on or inside a host organism at the expense of the host are called parasites; they do not normally kill their hosts. Parasitoids, however, are a unique type of parasite. They are of particular importance to farmers because they kill their pest insect hosts. Most parasitoids are wasps or flies that lay their eggs either on or inside the eggs, larva, or adult stage of their host. As the parasitoid larva develops, it feeds on and ultimately kills the host before emerging as a free-living adult outside the host.

Characteristics of parasitoids include:

- Most parasitoids are specialists, meaning they attack only one or a few host species.
- Many parasitoid wasps are extremely small, less than 1⁄25 inch (1 mm) in size, making them difficult to see with the unaided eye.
- Because of their small size, their presence is not obvious; you must carefully observe their hosts or use sticky traps to assess their populations in crop fields.
- Tiny parasitoids tend to be more susceptible to pesticides than predatory insects.

Pest Control and Farming

Beneficial insects are a natural part of all terrestrial ecosystems, and in the past were the primary means of pest control on farms. With the advent of chemical insecticides, contributions from beneficial insects often became forgotten. Rising insecticide use and reductions in natural habitat have led to declines of beneficial insect populations on many farms. Insecticides alone, however, have not solved the problem of crop pests. Despite their increasing use, both the absolute value of crop losses and the overall proportion of crop losses have increased in the past 40 years.

CONSIDERATIONS FOR HOME GARDENERS

Habitat enhancements for beneficial insects can be scaled down for yards and gardens, where beneficial insects can help suppress pests of fruits, vegetables, and ornamental plants.

For example, outbreaks of bagworms (*Thyridopteryx ephemeraeformis*), common pests that cause unsightly defoliation of ornamental shrubs and trees, often occur in simple landscapes dominated by turf grass. Researchers in Illinois experimented by increasing the number of wildflowers near bagworm-infested shrubs, and they found that rates of parasitism by beneficial insects were three times higher in bagworm populations near wildflowers than they were in bagworm populations surrounded by turf. Parasitoid wasps were more likely to seek their bagworm hosts in the presence of flowers.

Flowers are already a valued component of gardens and yards and, if selected and planted with beneficial insects in mind, can support predators and parasitoids that will reduce or eliminate the need for pesticide applications.

Secondary Pest Outbreaks

In some regions, inappropriate and excessive insecticide use has led to more pest outbreaks. These occur because insecticides often kill beneficial insects as well as pests, and populations of beneficial insects generally take longer to recover than pest populations. Furthermore, when insecticides kill off the first wave of pests and beneficial insects, a secondary pest outbreak — sometimes of a more damaging species than had previously been suppressed by beneficial insects — can result.

For example, *Lygus* bugs are key pests of cotton in California, and in late spring, when the bugs are active, farmers often

Parasitoid wasps are often very selective about their hosts.

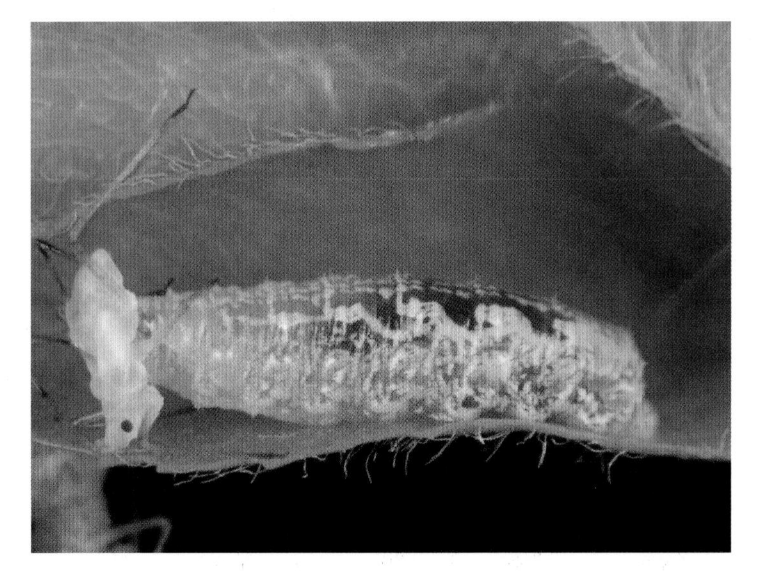

Flower fly larvae may consume up to 50 aphids per day.

apply broad-spectrum insecticides to control them. Armyworm caterpillars, aphids, and spider mites also feed on cotton. These herbivores in turn attract predators such as lady beetles, lacewings, big-eyed bugs, damsel bugs, and predatory mites, as well as parasitoid wasps. Following treatment with a broad-spectrum, nonselective insecticide for *Lygus* bug control, predator and parasitoid populations decrease and armyworm, spider mite, and aphid populations all increase. Researchers estimate that growers who treat cotton for *Lygus* bugs need to spend an additional 20 percent per growing season on insecticides to control secondary pest outbreaks.

Chemical Resistance

A predatory stink bug feeds on a plant-feeding stink bug.

Pesticide use can also lead to the development of chemical resistance in pests. When a few resistant individuals survive and breed, they create potential "super" pests that can no longer be effectively controlled by a particular chemical. At the same time, because beneficial insects typically produce fewer generations each year, they do not evolve a similar pesticide resistance. As a result, growers apply more and higher concentrations of specific pesticides to address the pest species' increased tolerance, and our beneficial insects suffer greater negative impacts. This cycle further undermines the ability of predators and parasitoids to do their work.

Pesticides will remain part of farming in the 21st century, but using new and rediscovered farming practices can promote beneficial insects that also contribute to pest control.

Enhancing Beneficial Insect Populations

Efforts to conserve native beneficial insects were unnecessary in the past, when crop fields were smaller and more diverse, and noncrop habitat more abundant. Research in the United Kingdom indicates that habitat enhancement through the creation of "beetle banks" (see chapter 10) and strips of wildflowers can increase beneficial insects and improve pest control. There is now limited but growing evidence in North America demonstrating the effectiveness of conservation biocontrol, and many farmers are now experimenting with ways to conserve and enhance native predators and parasitoids.

Extensive research now shows that farms with sufficient habitat demonstrate improved pest control from the beneficial insects supported by that habitat.

Solutions to Consider

You can enhance beneficial insect populations on your farm in a variety of ways, which we cover in detail in upcoming sections. Reducing pesticide use or switching to insecticides that target only specific pests are two ways to help conserve native beneficial insects.

You can also take more proactive steps. For example, as adults, many predators and parasitoids feed on wildflower nectar or pollen; when prey are lacking, nectar and pollen may be their primary food source. Therefore, one strategy to increase beneficial insect numbers is to enhance farm habitat with a variety of plants that bloom throughout the growing season. Such plantings can also provide alternate prey, egg-laying, and overwintering shelter for beneficial insects.

Approaches to Biocontrol

There are three approaches to managing crop pests using predators, parasitoids, and insect diseases: *Classical, Augmentative,* and *Conservation Biocontrol.* The focus of this book is on the last, but for context it is useful to understand the differences among these approaches.

Classical Biocontrol

Classical biocontrol is the release of a nonnative beneficial insect to control a typically nonnative pest. This process has been encouraged in the past when pests are accidentally introduced to a new location that lacks populations of their natural predators or parasitoids (natural enemies). Historically,

INTEGRATED PEST MANAGEMENT

Biocontrol strategies fall under a pest control framework called Integrated Pest Management (IPM). Pest control conducted using an IPM approach consists of four steps.

STEP 1. PREVENT CONDITIONS THAT FAVOR PESTS. This step includes practicing good sanitation and removing the pests' alternative plant hosts. It also incorporates strategies that interrupt pest cycles or keep their populations low through nontoxic means, such as managing habitat for beneficial insects, using mating-disruption pheromones, or practicing thoughtful crop rotation.

STEP 2. ESTABLISH AN ECONOMIC THRESHOLD. This is defined as a pest population density or level of crop damage at which action should be taken to avoid an economic loss.

STEP 3. MONITOR PEST POPULATIONS or pest damage.

STEP 4. TAKE ACTION TO CONTROL PESTS as a last resort only when populations or damage exceed the economic threshold; the cost of pest control should be less than the loss of income caused by the pests. Use pest control in the most targeted way possible, minimizing harm to beneficial insects and using the least toxic option for the situation.

Monitoring pest populations with insect traps (pictured here), sweep netting, and other techniques is a central part of IPM.

this process has been managed by government agencies and researchers working to identify beneficial insects within the pest's native range. Those beneficial insects are then screened for unintended environmental impacts, and ultimately if they are deemed "safe" to nonpest species, they are released into the new environment where their prey is already established.

Proponents of classical biocontrol consider this process to be environmentally safer than the use of insecticides to control new pests. Critics, however, point to examples of intentionally released "beneficial insects" that not only prey upon the intended pest, but go on to attack other, nonpest, species as well. Critics also point out that given enough time, native beneficial insects will adapt to prey upon newly introduced pests.

BENEFITS

Introducing a natural enemy from the native range of an introduced pest species has several advantages over other pest control practices, such as:

AN ALTERNATIVE TO INSECTICIDES. Classical biocontrol can significantly reduce the use of insecticides. Introducing natural enemies can, ideally, be less hazardous to both people and nonpest wildlife than are many insecticides.

EFFECTIVENESS. This approach provides a targeted attack on the pest in question and can lead to a long-term pest population reduction over a wide area.

LONG-TERM VALUE. Once introduced, the natural enemies often continue to be effective without further human intervention, and the benefit-to-cost ratio can be very high.

*The introduction of the vedalia beetle (*Rodolia cardinalis*), an Australian native, to California to control the invasive cottony cushion scale (*Icerya purchasi*) is a classical biocontrol success story.*

The introduction of the vedalia beetle (*Rodolia cardinalis*) to California in the late 1880s is widely viewed as a classical biocontrol success. Vedalia beetles, a type of lady beetle, were brought to California from their native Australia in an attempt to curb the cottony cushion scale (*Icerya purchasi*), a serious pest that threatened the growing California citrus industry. Vedalia beetles, which are host specific and attack only cottony cushion scales, knocked back pest populations quickly, and they continue to control it effectively today.

LIMITATIONS

The same biological traits of species introduced for classical biocontrol that are viewed as advantages, such as quick reproduction and the ability to disperse widely, also increase the chances of unexpected ecological interactions and effects.

DISRUPTION OF NATURAL SYSTEMS. Extreme caution must be used in classical biocontrol, because species introduced as biocontrol agents can disrupt native ecosystems. For example, a parasitoid tachinid fly (*Compsilura concinnata*) was introduced into North America beginning in the early 1900s to control the gypsy moth (*Lymantria dispar*), a nonnative pest of trees. While the gypsy moth has only one generation a year, the tachinid fly has several and therefore must attack nonpest insects to survive. The fly also has the ability to disperse to new areas quickly.

*Native giant silk moths such as the cecropia moth (*Hyalophora cecropia*) have undergone serious declines, in part due to parasitism by the parasitoid fly *Compsilura concinnata*, introduced as a classical biocontrol agent to control the gypsy moth.*

Initially scientists viewed the fly's dispersal ability as advantageous, with the expectation that it would spread to new areas and slow the gypsy moth invasion. Instead, the fly was found to attack more than 200 native insects in addition to the gypsy moth, including a number of native giant silkworm moths that subsequently underwent rapid population declines in the 1950s. The fly continues to be a persistent source of mortality for many butterflies and moths in the Northeast today and has spread across the continent.

The possibility of host switching, and the difficulty of predicting ecological interactions when introducing a species, are serious limitations of classical biocontrol.

IRREVERSIBLE IMPACTS. Introductions of species and their resulting interactions cannot be reversed.

Augmentative Biocontrol

When too few beneficial insects are present on the farm to keep pests below damaging levels, farmers sometimes use augmentative biocontrol to increase naturally occurring predator or parasitoid numbers. They purchase mass-reared beneficial insects and release them at a time when they can most effectively control pests. Often that release is earlier in the season than the naturally occurring beneficial insects would normally be present, or after their populations have been killed by pesticides, or on farms that do not have sufficient habitat to sustain beneficial insect populations. Many augmentative releases are performed with the expectation that they will not result in establishing permanent populations.

THE PROBLEM WITH RELEASING CONVERGENT LADY BEETLES

Convergent lady beetles (*Hippodamia convergens*), a species native to the United States, are commonly available for sale to farms, greenhouses, and gardeners. Rather than mass-rearing them in labs, some companies instead collect wild populations of the beetles from Western mountain valleys where they overwinter in large clusters. As a consequence, beetles removed from overwintering sites in places like the Sierra Nevada are sold and shipped nationwide.

This practice raises several ecological concerns among conservationists and researchers. In addition to removing large numbers of beetles from their natural environment (where other animals may depend upon them for food, and where the beetles may suppress local pest populations), there are additional risks associated with moving beetles around the country. For example, lady beetles collected from California and subsequently released in Eastern states may transfer pathogens or parasites to the local lady beetle populations. Interbreeding between geographically distinct populations may also be an issue: when the released lady beetles mate with local populations, it may negatively impact the local population by introducing traits that are adapted to a different environment.

Although the release of convergent lady beetles may work inside greenhouses, it is far less effective when beetles are released on farms or gardens because they tend to disperse immediately away from the site following their release. Due to their penchant for migration, the beetles often leave release sites without feeding or laying eggs.

Because there is little evidence of actual pest suppression, and because of the potential ecological risks involved, the Xerces Society does not recommend the release of lady beetles.

Convergent lady beetles overwinter in large groups. They are harvested from overwintering sites in one part of the country and sold in another for augmentative biocontrol.

Beneficial insect species that are commercially available may be native to the United States or have been introduced previously through classical biocontrol programs. Native species available for purchase for augmentation include the convergent lady beetle (*Hippodamia convergens*), green lacewing (*Chrysoperla* spp.), and insidious flower bug (*Orius insidiosus*). Introduced species include a predatory mite (*Phytoseiulus persimilis*) and parasitoid wasps (such as *Trichogramma brassicae*).

BENEFITS

Supplementing beneficial insect populations with mass-reared predators and parasitoids is sometimes viewed as a convenient and more ecologically sound substitute for insecticides. Specific benefits cited by advocates of augmentative biocontrol include:

PEST SPECIFIC. Augmentation can target and reduce a specific pest by releasing large numbers of natural enemies all at once.

AN ALTERNATIVE TO PESTICIDE. Augmentation efforts make use of insect species only, not pesticides.

PREVENTIVE. Reestablishing a desired beneficial species can also be useful as a preventive measure, providing control of pests in climates where the beneficial species cannot successfully overwinter.

IDEAL FOR GREENHOUSE USE. Augmentation can be very effective for use in greenhouses, where natural populations of beneficial insects may not be able to persist over time. For example, parasitoid wasps are routinely released in greenhouses to control whiteflies.

LIMITATIONS

Despite the benefits of augmentative biocontrol, real-world implementation of the practice has been limited by several factors:

A TEMPORARY MEASURE. Augmentation does not provide permanent or continuous pest suppression. While you can inundate your crop with large numbers of lacewings, for example, they will migrate to other areas if they cannot find supplementary food sources after they have controlled the pest population.

BEST IN CONFINED SPACES. This technique is most effective in an enclosed or semi-enclosed environment.

REQUIRES REPLENISHMENT. The grower must find a readily available commercial source of large numbers of beneficial insects and make continuous investments in them.

LIMITED APPLICABILITY. Augmentative biocontrol is not effective in all pest situations.

Conservation Biocontrol

Conservation biocontrol focuses on increasing the numbers and diversity of naturally occurring beneficial insects that are already present in the region. Predators and parasitoids need habitat to thrive. Because farm fields are subject to pesticide use, tillage, and mowing of ditches and field borders, they don't provide enough shelter, forage, and overwintering habitat for many beneficial insects. Conservation biocontrol focuses on providing food, shelter, or other habitat that has been demonstrated to increase numbers of beneficial insects.

Conservation biocontrol is a pest management practice that focuses on providing food and shelter for beneficial insects while at the same time reducing nonnative weeds and the pests that those weeds support. Simply put, conservation biocontrol is a merger between farming and ecology.

*Native sand wasps (*Bicyrtes quadrifasciatus*) build their solitary nests underground in sandy soils. Adult females provide for their young by hunting and subduing leaf-footed bugs and stink bugs, including the brown marmorated stink bug (*Halyomorpha halys*), an introduced, economically damaging pest.*

A female sand wasp brings a paralyzed stink bug back to her nest, where it will serve as food for her young. In her lifetime, one adult female sand wasp may collect about 50 brown marmorated stink bugs to feed her offspring.

For example, adults of many insect predators and parasitoids feed on pollen and nectar; therefore, patches of wildflowers in noncrop areas can help maintain (and even increase) their populations when prey insects are scarce.

BENEFITS

The benefits of conservation biocontrol are characterized by long-term impacts on the farm system. Achieving pest control benefits may require extensive planning and attention to detail, but once in place, a conservation biocontrol system will persist from year to year.

AN ALTERNATIVE TO INSECTICIDES. Conservation biocontrol can help reduce insecticide use.

SUPPORTS AN EXISTING ECOSYSTEM. By supporting natural enemies that are already present in the region, conservation biocontrol does not disrupt natural systems by introducing new species.

COMPLEMENTS OTHER STRATEGIES. Practices used in conservation biocontrol also benefit augmentative and classical biocontrol approaches.

BROAD BENEFITS POSSIBLE. In addition to pest population reductions, conservation biocontrol can benefit wildlife, such as pollinators and birds, as well as soil and water quality.

LIMITATIONS

Historically, intentional conservation biocontrol has been limited by a lack of information on how to restore and manage beneficial insect habitat. With this book and new informational resources being developed by researchers, conservationists, and crop consultants, we hope that limitation has largely been eliminated. Still, for some farmers, reasons cited for not adopting conservation biocontrol sometimes include:

LESS SPECIFIC THAN OTHER APPROACHES. Conservation biocontrol is not as targeted to specific pests as classical or augmentative biocontrol.

REQUIRES MORE TIME AND/OR LABOR. It requires attention to management of the noncropped land on your farm, or financial investment in the creation of new habitat.

DANGERS OF INTRODUCING EXOTIC BENEFICIAL INSECTS

As voracious predators of aphids and other insects, lady beetles have long been recognized for their potential to control certain pest species. The success of the vedalia beetle in controlling cottony cushion scale in the late 1880s inspired many subsequent introductions of lady beetles. In the years since, at least 179 species have been imported into North America intentionally, and several more species have arrived accidentally. There are now at least 27 species of nonnative lady beetles with established populations in North America. Unfortunately, not all of these introductions can be considered biocontrol success stories.

In contrast to vedalia beetles, which are specialized predators on cottony cushion scale, many of these other exotic lady beetles are generalist predators. Although their varied diet allows them to feed on many kinds of agricultural pests, these exotic lady beetles may also have a negative impact on insects that are not considered pests. Native, nonpest aphids may be impacted, as well as other potential prey. For example, both the seven-spotted lady beetle (*Coccinella septempunctata*), introduced from Europe, and the multicolored Asian lady beetle (*Harmonia axyridis*) are generalist predators that have been observed consuming butterfly caterpillars.

Exotic lady beetles may also impact native lady beetles. Both the seven-spotted lady beetle and the multicolored Asian lady beetle have significantly extended their range beyond their points of introduction, and are now the most abundant lady beetles in many areas. Several native lady beetle species have undergone severe population declines that parallel the population increases of their nonnative counterparts. There is evidence suggesting that seven-spotted lady beetles and multicolored Asian lady beetles are aggressive competitors for food and may be displacing native species through their superior abilities to find and consume prey. It has also been demonstrated that these beetles may also be impacting native lady beetles more directly, by preying on their eggs or larvae. While these exotic species alone may not be driving the declines of the native species, they are likely key contributors.

Additionally, multicolored Asian lady beetles are increasingly considered pests because they often seek winter shelter inside homes and also because they occasionally "nip" people. Multicolored Asian lady beetles also exude a foul-tasting substance from their joints when threatened, and this substance can leave stains when beetles are crushed. This has become an issue for vineyards: when the beetles are accidentally harvested along with clusters of grapes, they alter the palatability of wine.

*The introduced multicolored Asian lady beetle (*Harmonia axyridis*) may contribute to pest control but also negatively impacts native lady beetles and is considered a pest by homeowners and wine grape growers.*

Common Predators: Their Habitat and Prey

Beneficial Insects	Common Prey	Egg-Laying Sites
Ground Beetles Order: Coleoptera Family: Carabidae	Various soft-bodied pests including caterpillars, aphids, slugs, beetle larvae, insect eggs	In crevices or in the soil
Lady Beetles Order: Coleoptera Family: Coccinellidae	Aphids, scales, whiteflies, or mites Some also consume eggs or larvae of moths, beetles, flies, and thrips	On foliage near prey
Rove Beetles Order: Coleoptera Family: Staphylinidae	Mites, various insect eggs and larvae, and slugs	On leaves or under plant debris
Soldier Beetles Order: Coleoptera Family: Cantharidae	Aphids, slugs, insect eggs, and larvae (including caterpillars and other beetles)	In moist soil or leaf litter
Lacewings Order: Neuroptera Family: Chrysopidae Hemerobiidae	Aphids, thrips, mealybugs, whiteflies, caterpillars, and other soft-bodied insects	On foliage near prey
Damsel Bugs Order: Hemiptera Family: Nabidae	Aphids, leafhoppers, caterpillars, thrips, mites, various insect eggs	In plant tissue (leaf blade, petiole)
Big-Eyed Bugs Order: Hemiptera Family: Geocoridae	Aphids, scales, spider mites, lace bugs, thrips, whiteflies, small caterpillars, insect eggs	On leaves or in soil/duff near prey
Minute Pirate Bugs Order: Hemiptera Family: Anthocoridae	Thrips, aphids, mites, scales, psyllids Eggs of various insects, and small caterpillars (including corn earworm)	In plant tissue (leaf blade, petiole) or under bark
Assassin Bugs Order: Hemiptera Family: Reduviidae	Caterpillars, aphids, leafhoppers, grasshoppers, beetles, wasps, bees, flies, etc.	On plant leaves and branches
Flower Flies (Hover Flies) Order: Diptera Family: Syrphidae	Aphids, scales, mites, thrips	On foliage among prey
Predatory Wasps Order: Hymenoptera Families: Sphecidae Vespidae	Caterpillars, grasshoppers, crickets, beetles, spiders, treehoppers, aphids, true bugs, and flies	In solitary or social nests in the soil, in cavities in wood, or in nests made from resin, mud, or plant material

Shelter and Overwintering Needs	Supplementary Food Sources
Larvae or adults overwinter in grass clumps or dense woody vegetation When not hunting, adults take shelter under mulch, logs, brush piles, and stones	Omnivorous species may consume seeds, pollen, or detritus in addition to prey (Note: several species are also specialist feeders of weed seed such as lamb's-quarters and ragweed)
Adults overwinter in leaf litter, rock crevices, under bark, or inside buildings Adults shelter on undisturbed herbaceous and woody plants	Pollen Nectar Honeydew
Larvae, pupae, or adults overwinter under bark or vegetation Larvae and adults take shelter in leaf litter, grass thatch, logs, and brush piles	Many are omnivorous, and may eat fungal spores, pollen, or decaying organic matter
Larvae overwinter in leaf litter and loose soil Larvae and adults shelter underneath rocks, brush piles, or decaying wood	Pollen Nectar
Prepupae or adults overwinter in leaf litter, surface layer of soil, crevices, under bark, or inside outbuildings Adults benefit from sheltered field areas with minimal wind	Pollen Nectar Honeydew
Eggs or adults overwinter under leaf litter or grass thatch Nymphs and adults shelter in crop debris, mulch, or brush piles	Unknown
Adults overwinter in low-growing vegetation or under plant debris	Nectar
Adults overwinter under bark or leaf litter Nymphs or adults may shelter in plant stems, brush piles, and grass thatch	Pollen Nectar Plant sap
Eggs overwinter attached to plants Nymphs and adults overwinter under leaf litter, low-growing plants, and tree bark	Some drink nectar while waiting for prey on flowers
Prepupae, pupae, or adults overwinter in soil or leaf litter Adults prefer sheltered field areas with minimal wind	Pollen Nectar
Prepupae or adults overwinter inside nests, or as mated queens in leaf litter, outbuildings, or other sheltered areas	Nectar

Common Parasitoids: Their Habitat and Prey

Beneficial Insects	Common Prey	Egg-Laying Sites
Parasitoid Wasps Order: Hymenoptera Superfamilies: Ichneumonoidea Chalcidoidea	Many are host specific on various insects, including caterpillars, grasshoppers, aphids, sawflies, mealybugs, scales, whiteflies, beetles, etc.	On or inside host
Tachinid Flies Order: Diptera Family: Tachinidae	Some are generalist parasitoids of many insects Others specialize on caterpillars, grasshoppers, earwigs, sawflies, beetle larvae, or true bugs	On, inside, or near host

Other Common Beneficial Arthropods

Beneficial Insects	Common Prey	Egg-Laying Sites
Spiders Order: Araneae Family: Many, common examples include Araneidae Lycosidae Salticidae Linyphiidae	Many types of insects and arthropods	In egg sacs on webs, hidden in cracks on vegetation or rocks, or carried on the female spider, depending on the family
Harvestmen Order: Opiliones Superfamily: Phalangioidea	Many types of insects and arthropods	In soil or leaf litter
Predatory Mites Order: Acari Family: Phytoseiidae	Primarily spider mites, but also other mites, thrips, scales, mealybugs, psocids, whiteflies, small nematodes	On leaves or near prey colonies

Shelter Needs	Supplementary Food Sources
Prepupae, pupae, or adults overwinter within the host, or in soil or leaf litter, under bark or another protected place Adults prefer sheltered edges and areas with minimal wind	Nectar Honeydew
Larvae or pupae overwinter within their host, or pupae or adults overwinter in leaf litter or soil Adults prefer sheltered edges and areas with minimal wind	Nectar Honeydew

Shelter Needs	Supplementary Food Sources
Eggs or adults overwinter in silken nests in the soil, grass clumps, plant debris, under bark, or inside hollow stalks of vegetation Adults and spiderlings take refuge in leaf litter and mulch	A few species drink small amounts of nectar
Eggs overwinter in soil or leaf litter	A few species are omnivorous, and scavenge for dead insects
Adults overwinter on trees or in soil debris Shelter in humid microclimates in litter, on plants, or trees	Some may feed on pollen and nectar in the absence of prey

2

Why Farm with Native Beneficial Insects?

- Native beneficial insects annually provide $4.5 billion in pest control services to U.S. farms.
- Encouraging native beneficial insects can reduce or even eliminate the need for insecticides.
- Native insects offer huge benefits and few risks compared to exotic counterparts.
- Biocontrol also supports pollinators and other wildlife.

NATURAL HABITAT is essential to supporting beneficial insect populations, and the loss of such habitat has resulted in a loss of natural pest control services. Farms with natural or seminatural habitat around them tend to have more beneficial insects, less pest pressure, and less damage from pests. In general, farms with smaller field sizes, less pesticide use, and more noncrop habitat have the most beneficial insects.

Pest Control

While beneficial insects can maintain pest populations below economic thresholds, their services often go unnoticed by farmers and pest managers. Yet, when their populations are disrupted by loss of habitat or pesticide use, pest outbreaks may become more common.

Reduced Insecticide Use

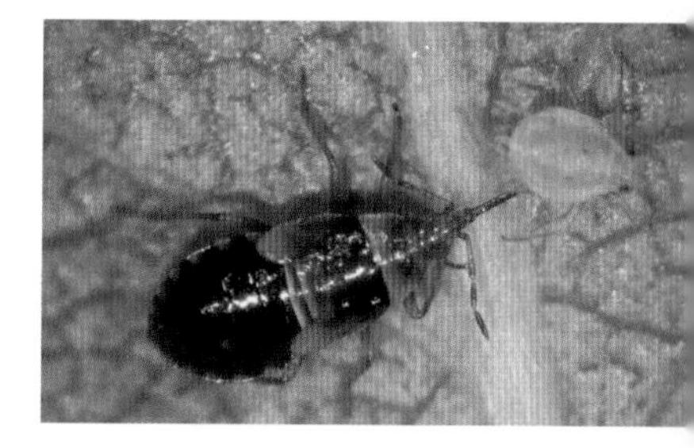

A pirate bug nymph feeds on an aphid.

Healthy beneficial insect populations can reduce the need for chemical insecticides. In order to capitalize on beneficial insect suppression of pest outbreaks, farmers must not only monitor pests, but also recognize the presence of beneficial insects. With close monitoring, they can reduce insecticide use and apply more targeted insecticides. This in turn will result in enhanced pest control in future years by allowing beneficial insect populations to build up and thrive.

Reduced Dependence on Nonnative Biocontrol Agents

The majority of beneficial insects in agriculture are native, naturally occurring species. In contrast, many pest species are exotic and were introduced by accident. Since they have left their natural enemies behind in their native range, these pests reproduce unchecked and cause economic losses in crops. Practitioners of classical biocontrol identify and make use of natural enemies that prey on the pest in its native habitat. Intentional release of these nonnative species can provide effective pest control, but the introduction of nonnative biocontrol agents may produce negative impacts on the natural ecosystem.

For example, a tachinid fly, *Compsilura concinnata,* was released repeatedly in North America from 1906 to 1986 as a biocontrol against several pests, including the introduced European gypsy moth (which defoliates forests throughout the East even today). Now the tachinid fly is implicated in the drastic decline of several nonpest silk moths in New England, including once-common species such as the cecropia moth (*Hyalophora cecropia*). Similarly, a recent study in Hawaii revealed that 83 percent of parasitoids found inside native moths were nonnative species that had originally been introduced for biological pest control.

Because of these and other unanticipated effects, classical biocontrol is controversial. If farmers and pest managers provide habitat for beneficial insects, there is less need for the release of nonnative insects. Supporting a diverse array of beneficial insects helps to keep native and nonnative pest populations below economically damaging levels. Also, given time, native beneficial insects typically adapt to prey upon new pest species.

WHAT IS CONSERVATION BIOCONTROL WORTH?

Scientists at Cornell University and the Xerces Society estimate the value of native predator and parasitoid insects for crop pest control in the United States to be at least $4.5 billion annually. This is a conservative estimate, based upon limited information. Those same scientists suspect the actual real-world value may ultimately be even higher.

A crop-specific estimation by Doug Landis from Michigan State University and his colleagues found that suppression of soybean aphid by beneficial insects was worth $239 million a year (based on 2008 soybean prices), in Iowa, Michigan, Minnesota, and Wisconsin alone.

A flower fly larva hunts soybean aphids, an invasive pest that causes significant damage to soybean crops in the United States.

Benefits beyond Pest Control

Conservation biocontrol can provide services to the farm in addition to pest control. For example, the same habitats that support predators and parasitoids also support a diverse array of crop pollinators, such as bumble bees and managed honey bees. These flowers also attract other colorful flower visitors, such as monarch butterflies. All of these, and other insects, in turn provide food for songbirds, game birds, and other wildlife.

Native grass and wildflower plantings for conservation biocontrol can also be incorporated into field buffer systems such as filter strips, grass waterways, roadside embankments, and septic drainage fields. These buffers reduce soil loss and improve water quality by filtering runoff from adjacent fields. They also help absorb and remove excess nutrients in farm systems.

Native shrub, grass, and wildflower plantings can also deprive some crop pests of habitat by reducing their host plants in noncrop areas. For example, in the Pacific Northwest, berry producers are increasingly concerned about the introduced spotted-wing drosophila (*Drosophila suzukii*), a fly that is a pest of soft fruit. The invasive Himalayan blackberry (*Rubus armeniacus*), a common weed along farm fencerows in the Northwest, serves as an alternate host for the fly, increasing its populations even when nearby crops are sprayed with insecticide to control it. To combat the fly, some growers are turning their attention to eliminating the Himalayan blackberry and replanting those fencerow areas with native shrubs that lack fruit on which the fly can reproduce.

Similarly, studies in the Central Valley of California clearly indicate that nonnative weedy plant species found along farm and road margins harbor more crop pests than do plantings of native species. Therefore, any efforts to replace Eurasian weeds with native habitats for beneficial insects can be doubly useful by both increasing predators and parasitoids and removing potential crop-pest habitat.

Benefits to Organic Farms

Habitat enhancements made on farms certified as organic receive additional benefits. The USDA National Organic Program mandates that organic farmers maintain or improve their on-farm natural resources, including soil, water, wetlands, woodlands, and wildlife. Over the last several years, advocates for wildlife and farm sustainability have worked with the National Organic Standards Board to develop mechanisms for organic farm certifiers to document that organic farms are maintaining this standard.

BENEFICIAL INSECTS IN THE HOME GARDEN

The value of encouraging beneficial insects in your yard or garden extends beyond natural pest control. Garden habitat for beneficial insects can be attractive, can reduce the need for pesticides, and can support charismatic wildlife like pollinators and birds. Plants that attract beneficial insects can also play multiple functional roles in your garden, such as helping to soak up excess rainwater or to return nutrients to the soil.

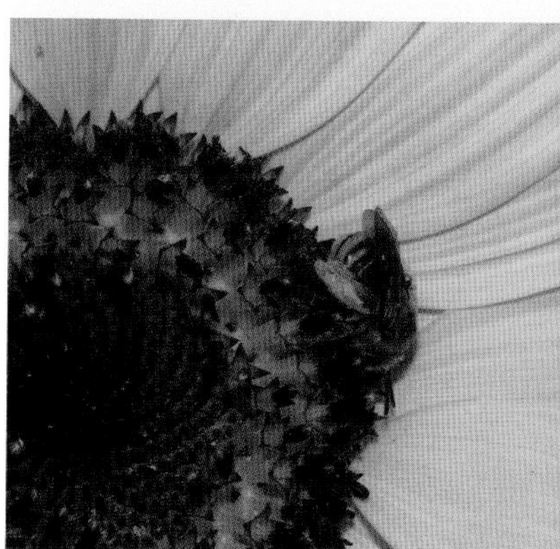

*Habitat for beneficial insects can support a variety of valuable wildlife, including pollinators such as this long-horned bee (*Melissodes *spp.).*

Beneficial insect habitat can help organic farmers fulfill the biodiversity conservation requirements of NOP certification.

Native wildflowers like those on this farm do not typically support pest insects.

BIOCONTROL, ORGANIC CERTIFICATION, AND THE BIODIVERSITY RULE

Conservation biocontrol can help farmers fulfill the biodiversity requirements for organic certification described in the National Organic Program (NOP) Rule. For example, the NOP definition of organic farming includes practices that:

"Foster cycling of resources, promote ecological balance, and conserve biodiversity." (§205.2)

"Maintain or improve the natural resources — the physical, hydrological, and biological features, including soil, water, wetlands, woodlands, and wildlife — of the operation." (§205.200 and §205.2)

Since the summer of 2012, organic farm certifiers have been required to check if farms are helping to secure on-farm biodiversity. Organic farmers who demonstrate that they are working to maintain, protect, or enhance habitat for beneficial insects on their land clearly are meeting these criteria.

CASE STUDY

Pest Management in Washington State Vineyards

FOR EASTERN WASHINGTON'S VINEYARDS in the 1990s, controlling grape pests meant using pesticides, and lots of them. At that time grape growers applied more than seven pounds of active ingredient of pesticide per acre per season, and the vineyards were alive with spray application machinery. Worse, most of the pesticides were broad-spectrum, as likely to kill the good insects and mites as the bad ones. The landscape was virtually devoid of life except the grapevines.

With the new century came a new awareness and desire by grape growers to reduce their dependence on pesticides and to produce fruit and wine containing as few chemical residues as possible. Washington State University scientists researched and developed biologically based Integrated Pest Management systems, and progressive growers put them into practice. Persistent, broad-spectrum pesticides were replaced by fewer applications of short-lived, narrow-spectrum "soft" chemicals identified as safe to key predators and parasitoids. Almost immediately, things began to change. Outbreaks of spider mites and mealybugs, common in the 1990s, became less frequent as natural enemies of these pests colonized vineyards.

Research on conservation biocontrol of grape pests intensified in multiple states with identification of key predators and parasitoids followed by a progressive understanding of their biology and ecology. For example, researchers found that the key natural enemies of grape leafhoppers, minute parasitic wasps in the genus *Anagrus*, overwinter within other leafhoppers living on rosaceous plants. Plantings of roses at the ends of grape rows or in specific areas within vineyards provided overwintering habitats for *Anagrus* hosts and hastened the migration of wasps into vines in the spring, improving biocontrol of leafhoppers.

Beyond just Washington, the integration of beneficial insect habitat into vineyards is an increasingly common practice throughout the U.S. wine industry.

Research on biocontrol of spider mites showed that improving the vineyard habitat by reducing the number of sulfur sprays used for disease control strengthened the diversity and abundance of predatory mites and suppressed spider mite populations. Other studies showed native bushes and trees acted as harbors for predatory mites, encouraging growers to set aside areas within vineyards for native plants.

Experiments with the use of nonnative flowering annuals as ground covers indicated that these ground covers improved biocontrol by attracting and retaining natural enemies of vineyard pests. Although it is difficult to sustain these nonnative plants in the dry environment of eastern Washington, the experiments stimulated contemporary research on native shrubs and bushes that could be used as beneficial ground covers.

Further improvements to vineyard habitats are under way, with native habitat restoration programs designed not only for predators and parasitoids but also for pollinators. The Washington grape industry is an excellent example of what can be achieved for agriculture, the local community, and the environment by farming for beneficial insects.

— DR. DAVID G. JAMES
Washington State University Department of Entomology

CONSERVATION BIOCONTROL ON ORGANIC FARMS

Recent studies show that pest control by naturally occurring predators and parasitoids on some organic farms may be as effective as pest control with pesticides on some conventional farms.

Beauty, Recreation, and Education

Finally, conservation biocontrol offers the additional benefit of beautifying landscapes. Having functional mass plantings of native wildflowers or flowering shrub hedgerows can light up a landscape. These habitats can provide recreational opportunities for farm families, such as high-quality habitat for hunting, bird watching, and learning about insects and wildlife. Attractive plantings along roadsides or market stands also support agritourism business models, especially if they include interpretive signs or other educational materials, and offer a way to connect with farm customers who are interested in sustainable agriculture issues.

CASE STUDY

Milkweed, Stink Bugs, and Georgia Cotton

THE FLOWERS OF MILKWEED (*Asclepias* spp.) can produce a rich supply of nectar. We wanted to find out the extent to which planting insecticide-free milkweed habitats in agricultural farmscapes could conserve bees and other insect pollinators, as well as enhance parasitism of insect pests.

In peanut-cotton farmscapes in Georgia, stink bugs, including the southern green stink bug (*Nezara viridula*) and the green stink bug (*Chinavia hilaris*), develop in peanut fields and then move on to feed on fruit in an adjacent cotton crop. We wanted to know whether strategic placement, in time and space, of a milkweed habitat between crops could lead to successfully increasing biocontrol of stink bugs in these agricultural settings.

In our experiments, we aimed to: (1) document feeding of stink bug parasites on milkweed nectar in the field; and (2) determine the impact of strategic placement of milkweed nectar provision on parasitism rates of stink bugs in cotton.

In this two-year study, we established plots with 25 milkweed plants per plot, and control plots without milkweed, along the edges of neighboring peanut and cotton fields. Weekly, throughout the growing season, we observed and recorded insects visiting flowering milkweed plants. We sampled cotton for stink bugs each week during the five-week period these pests colonized the crop.

Milkweed plants are very attractive to a wide range of beneficial insects, including solitary predatory wasps.

Over both years of the study, parasites and predators that attack the stink bugs (including both eggs and adults) readily fed on milkweed nectar in these insecticide-free habitats, as did insect pollinators, including honey bees, bumble bees, carpenter bees, leafcutter bees, and hover flies. For the first year of the study, parasitism of southern green stink bug adults by the stink bug parasitoid fly *Trichopoda pennipes* was close to five times higher in cotton plots with nearby milkweed habitat than in the control plots. In the second year of the study, combined parasitism of southern green stink bug, green stink bug, and leaf-footed bug (*Leptoglossus phyllopus*) adults by this parasite was at least three times greater in plots with nearby milkweed habitat than in control plots.

In only one year, there was indication that this management strategy alone can help maintain stink bugs below the economic damage threshold for this crop. Complementary management strategies, however, such as use of selective insecticides and trap cropping, may also need to be incorporated to suppress pests below economic thresholds throughout the growing season.

In conclusion, provision of a milkweed insectary habitat between peanut and cotton fields aided beneficial insect conservation and increased the rate of adult stink bug parasitism in cotton.

— DR. GLYNN TILLMAN, Crop Protection & Management Research Laboratory, U.S. Department of Agriculture Agricultural Research Service

3 Evaluating Beneficial Insect Habitat

- Like all animals, beneficial insects need food and shelter to survive.
- On farms with abundant natural habitat, wild insects can provide significant pest control.
- Farms with limited natural habitat can increase it with simple conservation features and ecological practices.

RECOGNIZING OPTIMAL HABITAT is the first step in conservation biocontrol. While insect identification is also important, it can be tricky to identify beneficial insects that closely resemble pests. Habitat identification, in contrast, is a relatively straightforward process that anyone can quickly learn.

Although it is useful, it is not always necessary to know all the species of beneficial insects on your farm. However, it is helpful to have a general understanding of their biology so that you can determine how best to support them. Begin by reviewing the table of common predator and parasitoid insects in chapter 1 and the identification guides in part 4, then use this chapter as a field guide to the landscape features where those insects are found.

Habitat Essentials

To contribute to pest control, beneficial insects must be present in large enough numbers in the crop, near the crop, or within the landscape to respond effectively to increasing pest densities. While crops often contain plenty of prey at certain times of the year, cropping systems alone cannot provide all the habitat beneficial insects need. You'll notice that several habitat requirements are shared by many of the predator and parasitoids described in this book, including:

- An abundance of flowers (especially native wildflowers) to provide pollen and nectar as a supplemental source of food
- Alternative prey when pest populations are low
- Relatively undisturbed shelter for reproduction and winter hibernation, such as bunch grasses, brush and rock piles, or logs.

No-till or reduced tillage systems can increase shelter for ground-dwelling predators like spiders.

Tillage can negatively affect predatory beetle larvae that reside in leaf litter or soil.

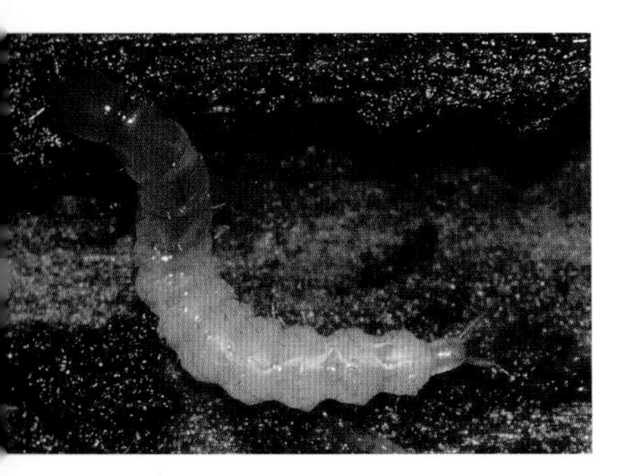

Cropping systems, which are short term and undergo frequent disturbances, cannot on their own support beneficial insects in large enough numbers to contribute to pest control. For example, tillage limits the suitability of crop fields as overwintering sites for ground beetles, spiders, and many other insect predators that take shelter in crop residue, leaf litter, or the top layer of soil. Tillage can also destroy the underground nests of predatory wasps. Within crop systems, beneficial insects also face an increased risk of exposure to pesticides.

And although crops may supply prey on which to feed, many beneficial insects need resources like nectar and pollen, or alternative prey. Beneficial insects need undisturbed, noncrop habitat nearby that can provide them with these additional resources and shelter if they are to survive, proliferate, and contribute to pest control.

Property boundaries and field edges can provide a suitable location for more permanent habitat features like hedgerows.

LADY BEETLE HABITAT USE

How a lady beetle uses various types of habitat and crops during the year.

In fall, lady beetles find habitat in which to overwinter as adults.

In spring, lady beetles feed on pollen and nectar.

Moving into the crop in search of prey, lady beetles reproduce there, laying eggs near prey.

When the crop is harvested, or prey becomes scarce, lady beetles move into nearby habitat for nonprey or alternative prey food.

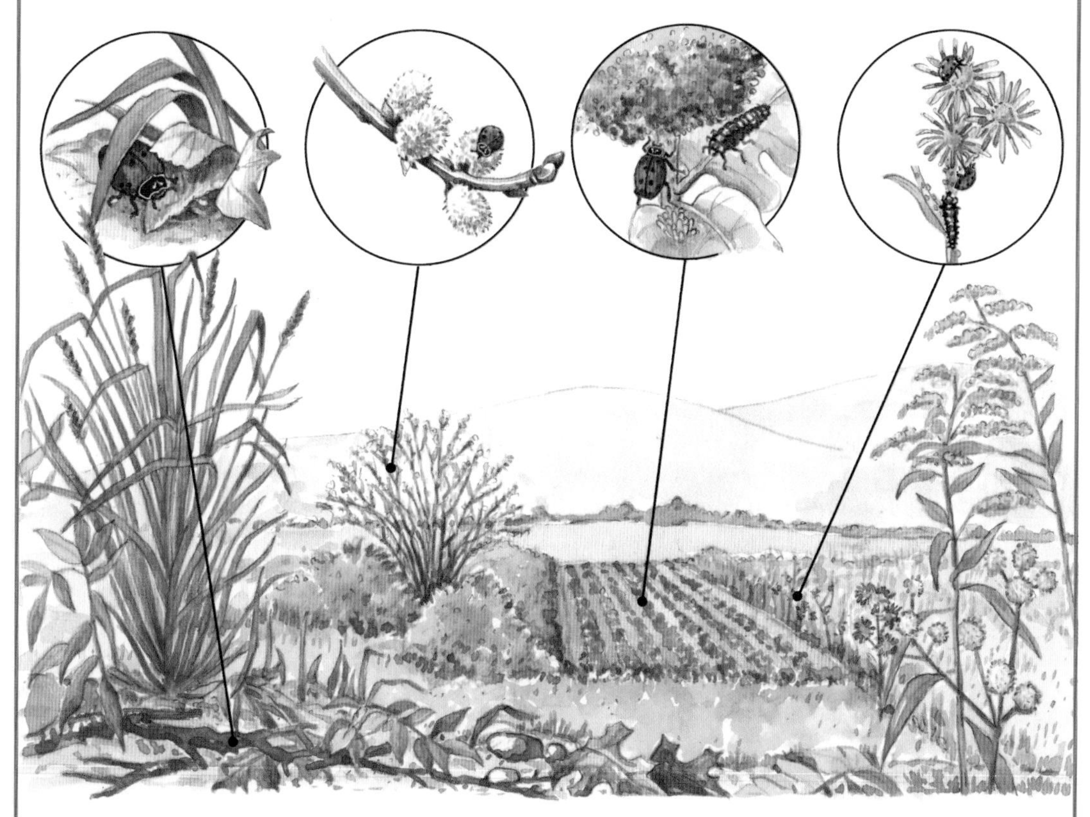

OVERWINTERING SHELTER: Leaf litter, crevices in bark, bunch grass thatch, rocks

CROP: Source of prey

NATIVE VEGETATION: Source of pollen, nectar, and alternative prey

Habitat in the Surrounding Landscape

Researchers have generally found more beneficial insects on farms within complex landscapes — landscapes that have some noncrop land — than on farms in a simpler, homogeneous landscape. Soybean fields in the Upper Midwest, for example, had more spider activity when located in a landscape surrounded by forests and prairies than in a landscape consisting entirely of other farm fields.

Not all beneficial insects respond similarly to the surrounding landscape, however; the ways they use habitat and the distance they can disperse to find new habitat determine their abundance. Species that are highly mobile are typically the first to colonize new habitat areas. For example, young linyphiid spiders can travel great distances on the wind (via small parachute-like silk strands) and are thus more able to spread far and wide when there are diverse natural areas near farms to serve as population sources. This means that even when farm habitat is less than optimal, nearby natural areas may continue to supply your field with spiders. In contrast, ground beetles travel limited distances and are more likely to be impacted (in good or bad ways) by farm habitat itself.

Even small remnant patches of natural habitat, such as native wildflowers growing along fencerows and roadsides, can help maintain stable long-term populations of beneficial insects.

Remnant and restored prairies of the Great Plains and Midwest, such as this site in Nebraska, are biodiversity "hot spots" that directly benefit the farms around them.

Habitat on Your Farm

You may already have habitat features on your farm that support beneficial insects, especially along roadsides, fencerows, ditches, field borders, and streams; around buildings; and in woodlots and natural areas. By looking at your whole property you may discover an abundance of beneficial insect habitat already present. Even if these features occur only in small, fragmented patches, taken together they can provide a mosaic of valuable habitats. Note, however, that it is important for these discrete patches to be located close to one another, as many insects may not travel more than a couple of hundred yards to hunt for food. It's also important that these habitat areas are located away from areas where pesticides are used.

CONSIDERATIONS FOR GARDENERS

Whether your garden consists of a small vegetable plot or multiple beds with fruits and vegetables, herbs, perennial wildflowers, or shrubs, you can integrate habitat for beneficial insects into your yard in a number of ways. Look for the Farm Practices Checklist starting on the next page. The same guidelines for augmenting beneficial insect habitat that apply to farmers can also be applied in the home landscape. Use the checklist to help you evaluate existing habitat and plan future changes to enhance that habitat.

Native plant field borders can reduce soil erosion from nearby agricultural fields while also supporting beneficial insects.

*While larvae of goldenrod solider beetles (*Chauliognathus pensylvanicus*) seek out prey like aphids or caterpillars, adult beetles are commonly found on flowers, where they feed on pollen and nectar.*

FLOWERS

Where native wildflowers are present, you can begin familiarizing yourself with some of the common predator and parasitoid insects in your area. Look for flowers that have a lot of insect visitors; on warm sunny days the best flowers will be especially active. For example, in many areas goldenrod is a very important source of late-season pollen and nectar for a huge variety of beneficial insects.

SHELTER

Predator and parasitoid insect species require hugely diverse kinds of shelter for reproduction and overwintering, so plan to provide as wide a variety of habitat features as you can. For example, certain wasp species nest underground in dry, sandy soils with sparse vegetation, while a number of predatory beetles overwinter in the thick crowns of native bunch grasses or lay their eggs underneath rotting logs. Consider undisturbed natural areas with patchy or overgrown vegetation, stumps, snags, rocks, or brush piles as potential refuge for beneficial insects.

Farm Practices Checklist

CROPPED AREAS

PESTICIDE PRACTICES	Current	Future
INTEGRATED PEST MANAGEMENT program that specifically addresses beneficial insect protection. [Note: If you do not use pesticides, skip this section and go to the Soil Practices section on page 37.] Pest management strategies are undertaken that include protection for beneficial insects. For more information, see chapter 11.	○	○
SELECTIVE INSECTICIDES are used, insecticides with little residual toxicity are used, or insecticides are applied at a time when beneficial insects are least vulnerable. Insecticides are selected or applied in order to reduce exposure to beneficial insects. For more informatio, see chapter 11.	○	○
SYSTEMIC INSECTICIDES are not applied to flowering crops before or during crop bloom. Beneficial insects can consume contaminated pollen or nectar if systemic insecticides are applied prior to or during bloom. For more information, see chapter 11.	○	○
A BUFFER EXISTS between insecticide applications to crops and noncrop habitat. A buffer, such as a windbreak, can reduce pesticide drift and can protect habitat. For more information, see chapter 11.	○	○

CROPPED AREAS *CONTINUED*

PESTICIDE PRACTICES *CONTINUED*	Current	Future
SPRAY DRIFT is carefully controlled. Reduce drift by spraying under proper weather conditions to avoid contaminating noncrop areas. For more information, see chapter 11.	○	○
SPRAY EQUIPMENT is calibrated annually. Settings of spray equipment are optimized for applications at low pressure, to deliver drops that travel less and reduce drift. For more information, see chapter 11.	○	○

SOIL PRACTICES	Current	Future
REDUCED OR NO-TILL SYSTEM. Tillage can reduce beneficial insect shelter, and can injure insects overwintering in the upper soil layer or in plant residue on the surface, as well as predatory wasps nesting in the ground.	○	○
USE OF COVER CROPS. Including cover crops in planting rotations helps to build soil health and can provide vegetative cover that shelters beneficial insects within crop fields. For more information, see chapter 8.	○	○
USE OF COVER CROPS THAT FLOWER (e.g., buckwheat). Flowering cover crops can support beneficial insects that consume pollen and nectar as part of their diet. For more information, see chapter 8.	○	○
VEGETATIVE COVER IN CROP ALLEYS OR FIELD ROADS (e.g., white clover in understory of perennial crops). Vegetative cover can add nutrients to the soil while providing shelter and resources for beneficial insects. For more information, see chapter 8.	○	○

NONCROPPED AREAS

ENHANCING FOOD RESOURCES AND MOVEMENT CORRIDORS	Current	Future
INSECTARY PLANTINGS. Insectary plantings can support beneficial insects within or directly adjacent to crop fields, can be easily managed, and the flowering can be timed to provide nectar and pollen when most needed to support key predators or parasitoids. For more information, see chapter 6.	○	○
NATIVE PLANT FIELD BORDERS can provide additional food sources for beneficial insects. For more information, see chapter 5.	○	○
HEDGEROWS can provide forage, alternative prey resources, and travel routes for beneficial insects. For more information, see chapter 7.	○	○
BUFFER STRIPS planted with native grasses and wildflowers can provide a variety of resources for beneficial insects. For more information, see chapter 9.	○	○

Farm Practices Checklist continued

NONCROPPED AREAS *CONTINUED*

ENHANCING FOOD RESOURCES AND MOVEMENT CORRIDORS *CONT.*	Current	Future
MULTIPLE SPECIES of blooming trees, shrubs, or wildflowers present in each season ensure there are plants flowering from spring to fall. Sequential blooming wildflowers, shrubs, or trees can provide resources for beneficial insects throughout the growing season. For more information, see chapter 4.	○	○
PLACE PLANTINGS around farm to connect landscape elements (corridors). Linking habitat patches or including linear plantings can help beneficial insects travel through the farm landscape. For more information, see chapter 4.	○	○

ENHANCING REFUGES, OVERWINTERING, NESTING	Current	Future
COVER CROPS in field margins or in the understory of perennial crops help build soil health and can provide vegetative cover that shelters beneficial insects. For more information, see chapter 8.	○	○
PERMANENT PLANTINGS WITH NATIVE BUNCH GRASSES such as field margins can provide undisturbed shelter or overwintering sites for beneficial insects. For more information, see chapter 5.	○	○
PERMANENT PLANTINGS WITH TREES OR SHRUBS (e.g., hedgerows, riparian areas, forest edges) can provide undisturbed shelter or overwintering sites for beneficial insects. For more information, see chapters 7 and 10.	○	○
BUFFER STRIPS planted with native grasses and wildflowers can provide undisturbed shelter or overwintering sites for beneficial insects. For more information, see chapter 9.	○	○
AREAS OF WELL-DRAINED SOIL with patches of bare ground that are not tilled (well-drained sandy soils with sparse vegetation) can be nesting sites for ground-nesting predatory wasps.	○	○
CLEAN SOURCE OF WATER. An uncontaminated water source is important for predatory wasps during their nest-building, and may be used by other beneficial insects.	○	○
BEETLE BANKS can provide undisturbed shelter or overwintering sites for ground beetles and other beneficial insects. For more information, see chapter 10.	○	○
ARTIFICIAL NEST BLOCKS may be utilized by solitary predatory wasps that nest in tunnels. For more information, see chapter 10.	○	○
BRUSH PILES can be shelter or overwintering sites for some beneficial insects, and snags may provide nesting sites for tunnel-nesting predatory wasps. For more information, see chapter 10.	○	○
INSECT HOTELS may provide shelter or overwintering sites for some beneficial insects. For more information, see chapter 10.	○	○

NONCROPPED AREAS *CONTINUED*

LONG-TERM MAINTENANCE OF NONCROPPED AREAS	Current	Future
AREA THAT IS BURNED, MOWED, OR HAYED in a rotation of parcels, with ⅓ or less of the habitat disturbed each year, means beneficial insect populations are less impacted because only a fraction of available habitat is disturbed at a time. For more information, see chapter 12.	○	○
ROTATIONAL GRAZING or selective grazing can be used to promote wildflowers, which in turn can support beneficial insects. For more information, see chapter 12.	○	○

FEATURES OF SURROUNDING LANDSCAPE	Current	Future
THE LANDSCAPE SURROUNDING YOUR FARM contains some natural or semi-natural vegetation (e.g., prairie or grasslands, chaparral, woodlands, riparian habitat, wetlands). Landscape composition can influence the abundance of some groups of beneficial insects. Where natural habitat in the surrounding landscape is absent, working with local conservation agencies might provide opportunities for local habitat restoration.	○	○
IF NEIGHBORING LAND IS CROPPED, windbreaks or buffers can be planted to reduce drift between farms. A buffer such as a windbreak can reduce pesticide drift and can protect habitat. For more information, see chapter 11.	○	○

CASE STUDY

Beneficial Insects Save Christmas

THE ECOLOGICAL PRINCIPLE underpinning conservation biocontrol states that plant diversity drives insect diversity. In practical terms, this means that more plant species in an area support greater populations of insects, spiders, mites, and so on. While this pattern, and the potential pest problems, may concern some growers, in fact greater insect diversity typically leads to improved control of plant-feeding insects by predators and parasitoids. Increasing diversity tips the balance of power away from plant feeders and toward beneficial species that can help control insect pests.

This is the main reason that natural ecosystems (e.g., tallgrass prairies) rarely see pest outbreaks: a diverse group of predators and parasites responds quickly to growing pest populations to quash potential outbreaks. This same diversity is not available in most agricultural monocultures.

A goal of ecologically inspired agriculture is to help agroecosystems more closely resemble natural ecosystems whenever possible, which will in turn facilitate conservation biocontrol. This goal is readily achieved in perennial cropping systems like hayfields, orchards, vineyards, and plantations. In orchards and vineyards, for example, growers can promote populations of natural enemies by sowing flowering plant species in the rows between the vines or trees.

In Christmas tree production, many growers face pine needle scale (*Chionaspis pinifoliae*). One of the most serious pests of ornamental pines in the United States, this scale species has been labeled the "white malady" because heavily infested trees appear whitewashed. Insecticide-based management programs combating pine needle scale often exacerbate scale populations because insecticides are harder on natural enemies than on the scale population itself. To better control pine needle scale in Illinois, we devised a simple "conservation biocontrol-inspired" management plan: mow the grass between the rows less often. Less frequent mowing, or even mowing alternate rows, allowed resident weedy species like Queen Anne's lace (*Daucus carota*), chicory (*Cichorium intybus*), and white clover (*Trifolium repens*) to proliferate and flower. This provided habitat, floral resources, and alternative prey to a diversity of predators, including lady beetles, lacewing larvae, harvestmen, and tiny parasitoid wasps that can effectively control the scale population.

This simple solution to an insect pest problem worked quite well for our growers and also appeared to decrease the incidence of insect pest populations in general. The program did force growers to be a bit more diligent about trimming weeds around each tree, because mice like the weedy rows too and can girdle young saplings. Nevertheless, controlling a few weeds has been a small price to pay for less insecticide use and more reliable insect control.

— DR. JOHN F. TOOKER, Department of Entomology
Pennsylvania State University

Christmas tree habitat.

PART 2

Improving Beneficial Insect Habitat

WHATEVER ITS CURRENT CONDITION, every farm offers unique opportunities to further improve its beneficial insect habitat. In this section we explore the basic design process of adding new habitat features to the farm. Permanent additions include conservation buffers, native wildflower meadows, and hedgerows; temporary elements include cover crops and annual insectary strips.

With most of these features, the first step of the design process is to select the best plant species for the project. That is where we begin this section.

Restored prairie and meadow areas within a farm provide food and shelter for beneficial insects, and in turn they contribute to pest control in adjacent crops.

4 Designing New Beneficial Insect Habitat

- Native plants are the cornerstone of conservation biocontrol.
- Specific wildflowers can help maximize beneficial insect abundance.
- Brush piles and other features provide places for insects to lay eggs and overwinter.
- Native plants do not typically support pest populations.

EXPANDING FARM HABITAT for beneficial insects is key to conservation biocontrol. Increasing the availability of flowers — especially native wildflowers — is often the single most important strategy for increasing the abundance and diversity of beneficial insects. Like pollinators, such as bees and butterflies, many insect predators and parasitoids feed on flower nectar or pollen during one or more of their life stages.

Flower flies, for example, typically feed on aphids and other small insects during their larval stage. As adults, however, they feed on flower nectar or occasionally pollen. Similarly, parasitoid wasps typically feed on flower nectar as adults while various predatory insects feed on pollen as a supplemental source of protein when prey is in short supply.

By increasing the availability of flowers, you increase the abundance, longevity, and reproductive potential of the beneficial insects on your land. While integrating habitat like wildflowers into a farm system can seem daunting at first, this chapter and the others that follow provide examples that can be surprisingly easy to integrate.

What Beneficial Insects Need

The specific needs of beneficial insects mirror those of other wildlife, and indeed the needs of other animals in general: food, shelter, and protection from threats such as pesticide use. The basic building blocks needed to provide these requirements are typically nothing more than natural materials like native plants, brush piles, stumps, snags, and piles of stones.

Native Plants

Where available and affordable, native plants should be prioritized in conservation biocontrol efforts. The logic behind this emphasis on native plants is that native beneficial insects have adapted over thousands of years to live in close association with native plant species. For example, research across the country now routinely demonstrates that native plants support many more native insects than do nonnative plants.

Adult flower flies need nectar and occasionally pollen from flowers in order to survive and reproduce.

In recent years, researchers like Rufus Isaacs and Doug Landis from Michigan State University observed that a greater abundance of beneficial insects were associated with native perennial wildflowers than with nonnative plant species. Similarly, Doug Tallamy at the University of Delaware estimates that native plants may support as many as 35 times more caterpillar biomass than do nonnative plants. With few exceptions, native caterpillars supported by native plants are not crop pests, and can serve as alternative prey or hosts for beneficial insects. For example, the great ash sphinx moth caterpillar (*Sphinx chersis*) feeds on native ash trees, and the snowberry clearwing moth caterpillar (*Hemaris diffinis*) on snowberry and dogbane. Both moths are alternate hosts for parasitoid wasps that also attack tomato hornworms.

Where native plants are not affordable, or are only available in limited quantities, some noninvasive nonnative flowers, such as cover crops of annual buckwheat, can help supplement the habitat requirements of beneficial insects.

AVOIDING NUISANCE PLANTS

Another thing to consider when you are creating new habitat for conservation biocontrol is to avoid planting species that harbor pest insects, support crop diseases, or are potentially weedy. In most instances, crop pests prefer crops or nonnative weeds to native plants. Thus, providing a variety of native grasses, wildflowers, or shrubs typically benefits predators and parasitoids over crop pests. Note that in a few cases native plants may share the same pests and diseases as crop plants; some common examples of those shared pests and diseases

PREDATOR COMMUNITY INTERACTIONS

Habitat diversity on farms can increase the complexity of food webs. Enhancing the diversity of predators and parasitoids can improve natural pest control, but less predictable interactions can also result. For example, alternative prey can sometimes be more attractive to predators than crop pests. Similarly, predators can prey on other predators, or become hosts to parasitoids.

Wolf spiders, for example, may prey on damsel bugs that prey on squash bug nymphs, and so indirectly reduce natural pest control to a squash crop. Some parasitoids, known as hyperparasitoids, are parasitoids of other parasitoids. Cotesia wasps (*Cotesia congregata*) that parasitize hornworm caterpillars (*Manduca* spp.) are themselves parasitized by a cotesia hyperparasitoid (*Hypopteromalus tabacum*).

Sometimes the roles insects play in a food web and their contribution to pest control can be location and context dependent. If plenty of large prey sources, such as slugs, are available on mixed vegetable farms, large ground beetles tend not to prey upon the smaller rove beetles (a predator of pest fly eggs). However, when large prey sources are absent and large ground beetles do prey on rove beetles, pest flies increase, because there are fewer rove beetles to consume their eggs. Examples like these are relatively uncommon, however, and the benefits of habitat on farms for natural pest control generally outweigh the downsides.

are listed in the sidebar later in this chapter. Fortunately, such examples are not very common for most crops.

While in most instances it's best to avoid planting alternate pest host plants, a nuanced and alternative perspective promoted by some researchers recommends purposefully planting such species, known as **banker plants**, within greenhouses, high tunnels, or adjacent to farm fields. Banker plants lure pests away from the crop and thus support a stable population of beneficial insects (usually parasitoids) that prey upon them. The banker plant concept assumes that those parasitoids will then produce surplus numbers that migrate into crops.

A similar but less risky banker plant strategy is to plant species that support noncrop pests, which can serve as alternate prey or hosts of the same beneficial insects that control the target crop pest. According to findings from multiple research groups in Europe and Asia, the cotton aphid (*Aphis gossypii*), a very common and damaging pest of greenhouse vegetables, is better controlled by a parasitoid wasp (*Aphidius colemani*) when wheat or barley plants are included inside greenhouses. The grasses, acting as banker plants, support wheat and barley aphid pests (*Rhopalosiphum padi*), which serve as an alternate host for the parasitoid wasps.

The banker plant concept is likely to be most effective with parasitoids because they require specific prey (e.g., aphids), while generalist predators attack many types of insects and can occupy many types of habitat. When more research has been conducted and plant selection can be guided more carefully, banker plants may be effective for conservation biocontrol, but for now banker plants are more often used in combination with augmentation biocontrol inside greenhouses or other confined spaces.

GETTING HELP WITH CONSERVATION EFFORTS

United States Department of Agriculture
Natural Resources Conservation Service

The USDA Natural Resources Conservation Service (NRCS) provides financial and technical assistance to support conservation efforts for beneficial insects and other wildlife on farms. Conservation programs, such as the Environmental Quality Incentives Program (EQIP) and others, can be used to help farmers establish and maintain beneficial insect habitats. These habitats include native plant hedgerows, flowering cover crops, field borders, grassed waterways, and more. For information on farm conservation programs, contact your nearest NRCS office at www.nrcs.usda.gov.

Diverse Flower Species

Beneficial insects need the nectar and pollen that flowers provide. Because predator and parasitoid species are active throughout the growing season, it is best to plant a variety of flower species that bloom in succession throughout the year. Diverse flower species will attract additional beneficial insects while also supporting those already in the area. If flower availability is low, many predators and parasitoids will leave a farm and migrate to new areas in search of flowers.

More than simply supplying valuable pollen and nectar to beneficial insects, flowers also attract alternative prey to help sustain predators and parasitoids when crop pest populations are low. Thus a succession of blooming flowers on the farm throughout the growing season provides a stable source of food that beneficial insects need in order to thrive as their prey populations rise and fall.

Some of the best plants for conservation biocontrol are included in seed mix tables in chapter 5, and a more comprehensive guide can be found in part 5 of this book.

Alternative Prey or Hosts

*This assassin bug (*Apiomerus crassipes*) has captured a plant-feeding stink bug (*Cosmopepla lintneriana*).*

This lady beetle larva is feeding on milkweed aphids, a noncrop pest and an alternative source of prey that will tide the beetle over until crop pests arrive.

Alternative prey or hosts in nearby habitat can support predators or parasitoids through fluctuations in crop pest populations, including during seasons when a pest's particular life stage cannot be attacked by beneficial insects.

Predators can build their populations early in the growing season, before pests arrive, by feeding on alternative prey, and will be able to move into a crop to provide pest control as soon as pests arrive. For example, sheet-weaving spiders (Linyphiidae) are important predators of cereal aphids in winter wheat fields. In the absence of aphids, however, the spiders use springtails (Collembola; a group of small plant-litter decomposers) as an alternative source of prey. Farm practices, such as reducing tillage and establishing permanent native plant field borders, help support more springtails and, in turn, more spiders.

Though parasitoids are generally much more host specific than are predators, some parasitoids can utilize multiple hosts, and may in fact need multiple hosts in order to persist in some crops. In California, tiny chalcid wasps known as fairyflies (*Anagrus* spp.) are parasitoids that attack the eggs of the leafhopper (*Erythroneura elegantula*), a pest of grapes. However, the wasps need to overwinter outside of the vineyard on alternate hosts, because the grape leafhopper overwinters as an adult rather than as an egg. Blackberry leafhoppers (*Dikrella cruentata*) feed on wild blackberries (*Rubus* spp.) as well as the cultivated prune (*Prunus domestica*) and overwinter as eggs. Blackberry leafhoppers can serve as alternate hosts for fairyflies, if *Rubus* spp. grow naturally or are planted nearby. Fairyflies can contribute to control of the grape leafhopper if habitat for the alternate host is provided.

The best habitat for insect predators and parasitoids consists of diverse and abundant wildflowers that provide pollen and nectar throughout the growing season.

FOOD SUPPLEMENTS FOR BENEFICIAL INSECTS?

Nonprey foods such as pollen, nectar, extrafloral nectar, and honeydew are important for certain life stages of many beneficial insects. These foods enhance longevity, reproduction, and the ability to hunt prey. While farm habitat can provide these types of nonprey foods, another approach is the use of food sprays. Food sprays are artificial supplements that typically contain sugar and yeast-based proteins that, when applied to crops, are intended to mimic the nutrition provided by natural food sources such as pollen or nectar, and thus attract beneficial insects. To date, however, evidence suggests that food sprays are inconsistent in their effectiveness, depending upon the crop and product being applied.

Published research indicates that food sprays can increase the abundance of some groups of beneficial insects, which in turn can decrease some pest populations and crop damage. There is also evidence to suggest that food sprays may be used to reduce insecticide exposure by aggregating predators like lady beetles in one field while another is being treated with insecticides.

Despite these promising applications, there are also notable limitations. In some situations where food sprays were used, pests increased rather than declined. Some researchers also reported that sprays were toxic to certain crops or that crops were damaged by fungal disease or experienced stunted growth. In addition, food sprays are short-lived under field conditions, and so cannot provide nonprey food resources in the long term. The response of different crops and beneficial insects is variable, so food sprays will need to be evaluated on a case-by-case basis before they can become widely adopted.

Permanent, noncultivated grass strips between crop fields can provide winter cover for beneficial insects and other wildlife. Adding wildflowers, shrubs, and more diverse vegetative structure to a strip like this would increase its value for pest control even more.

Shelter

Finally, in addition to nectar, pollen, and alternate sources of prey or hosts, beneficial insects need shelter to survive winter or periods of bad weather, to transform between their larval and adult stages, and to lay their eggs. We don't fully understand many species' preferred types of shelter, although it is generally accepted that increasing the structural diversity and stability of a habitat over time can increase the abundance of beneficial insects. For those insects whose preferred overwintering sites are well understood, we can take steps to create specific habitat. For instance, we know that predatory ground beetles prefer to overwinter in bunch grasses on elevated ground. **Beetle banks** (described in chapter 10) can provide this undisturbed winter cover.

If your farm lacks structural diversity, you can take concrete steps to add shelter. Planting fallow ground with native bunch grasses is one simple strategy that can provide overwintering shelter for various predatory beetles. In California, for example, lady beetles are known to cluster inside clumps of native bunch grasses during the winter. Patches of bare ground between bunch grass clumps provide direct access to the soil for predatory ground-nesting wasps. Wood-nesting solitary wasps, on the other hand, can be readily attracted to bundles of hollow reed, bamboo, or cupplant stems, or drilled wood nest blocks like those used to attract mason bees. Such artificial nests can be hung like birdhouses around farm fields (see chapter 10 for more information).

Where there is a lack of adequate egg-laying and overwintering sites, artificial structures like this can be created.

ADJACENT HABITAT AND PEST POPULATIONS

Farmers are sometimes concerned that habitat for beneficial insects may increase pest populations. So do restored areas increase weeds or insect pests in adjacent crops?

WEEDS The native plants recommended in this book are not weedy and in most cases are unlikely to invade other areas. If you are careful about plant selection, seeding densities, and spacing, you may find that native plants eventually crowd out weeds. This can reduce weed invasions into your fields. Over the long term, areas restored with native wildflowers generally need minimal upkeep, although weeds may eventually reinvade the area and require some maintenance. Using native shrubs for habitat has the benefit of lasting for decades. For the first three to four years after planting native shrubs, you'll have to control weeds manually or with herbicides between plantings. However, as shrubs mature and fill gaps, weeds are increasingly excluded. To reduce the need for weed control while your shrubs are maturing, you can sow native wildflowers and grasses in between immature shrubs, or use mulch.

PEST INSECTS Multiyear studies of mature, native plant hedgerows in Yolo County, California, show that pest insect populations are lower in hedgerows than they are in weedy field edges. In addition, the ratio of beneficial to pest insects is much greater in native plant hedgerows than in weedy edges, meaning there are more beneficials per pest insect when native plants are present. Conversely, nonnative weeds such as Himalayan blackberry, mallow, and wild mustard are known to harbor pest insects. For example, Himalayan blackberry canes that bend over and touch the ground provide shelter for stink bugs during the winter. Removing Himalayan blackberry and replacing it with native plants can reduce overwintering stink bug populations and infestation in adjacent crops. Aphid populations are also much lower on native vegetation than they are on invasive weeds such as wild mustard. Studies also show that there are fewer pests and more beneficial insects close to a native hedgerow than there are in weedy edges.

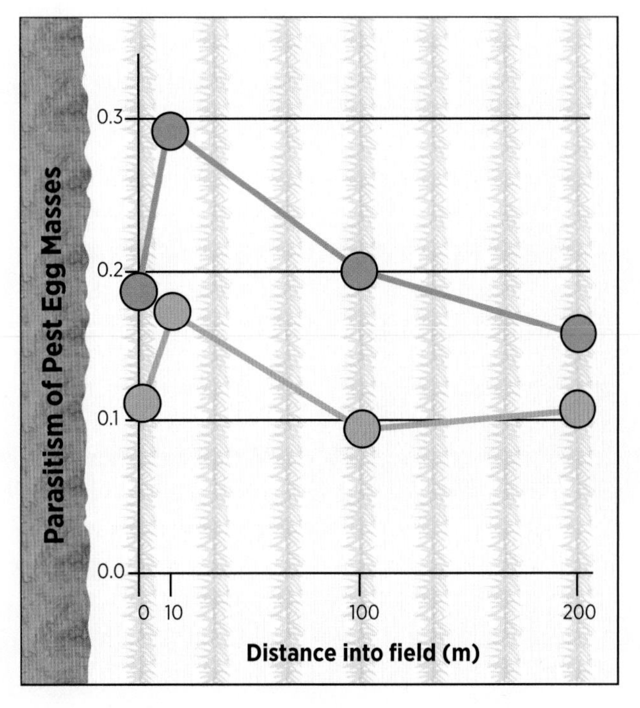

Mean proportion of parasitized stink bug egg masses over two years (2009–10) in tomato fields with and without hedgerows. There was greater parasitism on farms with hedgerows at 10 and 100 meters into the crop fields.

Parasitism on farms with hedgerows

Parasitism on farms with weedy field borders

Morandin, L., R. Long, and C. Kremen. 2014. Hedgerows enhance beneficial insects on adjacent tomato fields in an intensive agricultural landscape. Agriculture, Ecosystems & Environment.

Multiyear studies of native plant hedgerows in California show that pest populations are lower in hedgerows than they are in highly disturbed field edges dominated by nonnative weeds.

*Flower shape can influence the types of insects that are attracted to a particular plant. Beneficial soldier beetles, for example, are often attracted to flowers with lots of perching space like this cupplant (*Silphium perfoliatum*).*

Habitat Size and Location

How large and how close to crops do natural habitats need to be in order to provide pest control services? While these and other questions are still being explored by researchers, we do know that crop fields with a lot of noncrop habitat around them tend to have greater beneficial insect populations and lower pest pressure. Also, smaller fields interspersed with natural areas allow beneficial insects to move deeper into crop fields. Because beneficial insects may not travel as far as pests, it is important to create habitat on as many border areas as possible.

Field borders a few feet wide, running the length of the field, have been shown to increase beneficial insects in adjacent fields. Edge habitats, although not large, will increase predation and parasitism of pests within crops.

Habitat should be created as close to the target crop as possible, as long as it remains safe from pesticide drift. Although many predators can travel a considerable distance to colonize agricultural fields, some will not go far; for example, most parasitic wasps may only travel a few hundred feet.

PESTS AND PARASITOIDS IN NATIVE HEDGEROWS VS WEEDY FIELD EDGES

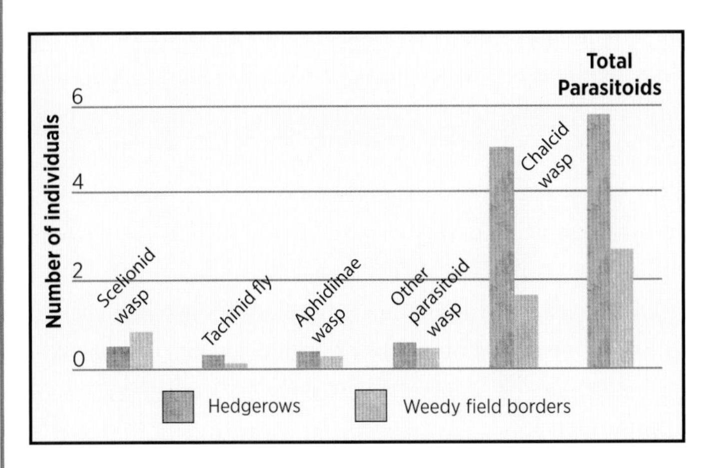

Mean abundance of parasitoids (left) and pests (below) in non-native weedy field borders (brown) versus field borders consisting of native plant hedgerows (green) over two years in sweep net samples.

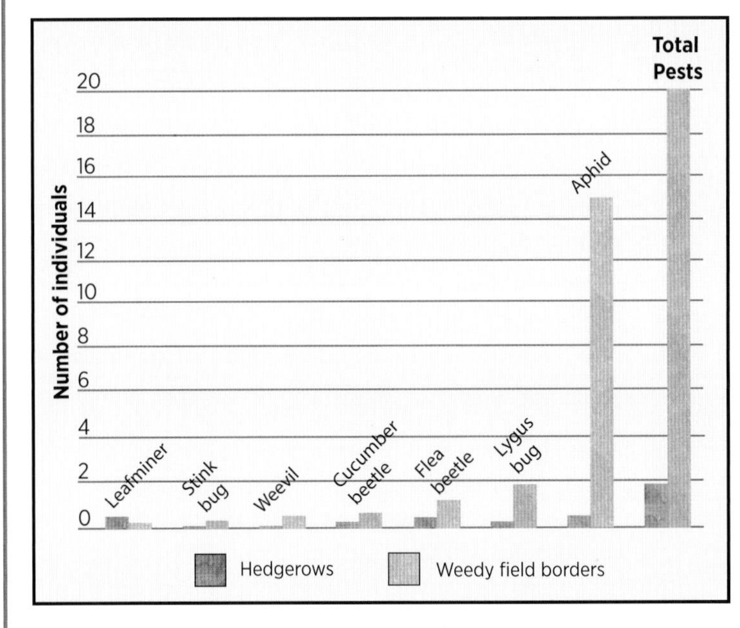

Morandin, L., R. Long, and C. Kremen. 2014. Hedgerows enhance beneficial insects on adjacent tomato fields in an intensive agricultural landscape. Agriculture, Ecosystems & Environment.

Wildflower Selection

DIVERSITY IS A CRITICAL FACTOR when considering wildflowers to plant for beneficial insects. Natural flower-rich habitats, such as remnant Midwest tallgrass prairies, may include dozens of wildflower species, but for most conservation areas on farms, even as few as 10 carefully chosen plant species will provide a good foundation for attracting a wide variety of beneficial insects.

*Members of the carrot family, like golden Alexanders (*Zizia aurea*), have shallow flowers with readily accessible nectar.*

*Some plants, like partridge pea (*Chamaecrista fasciculata*), produce nectar in extrafloral nectaries at the base of the leaves.*

One way to begin selecting flowers to plant is to think about the physical characteristics of various predators and parasitoids that use pollen and nectar resources. Wasps, for example, generally have short tongues and are unable to reach nectar hidden in the deep reservoirs of some flowers. Consequently, nectar-collecting wasps are most frequently observed on shallow flowers with readily accessible nectar, such as goldenrods, or members of the carrot family such as golden Alexanders (*Zizia aurea*) or rattlesnake master (*Eryngium yuccifolium*). Wasps also frequently seek out sugar sources other than nectar, such as honeydew (the sugary excrement of aphids and scale insects), rotting fruit, or the extrafloral nectaries of plants like partridge pea (*Chamaecrista fasciculata*), which secretes nectar at the base of leaves.

Beneficial flies also have short mouthparts that are best suited to open or small flowers such as yarrow (*Achillea millefolium*), fennel (*Foeniculum vulgare*), alyssum (*Lobularia maritima*), or fleabane (*Erigeron* spp.). Beetles often feed where the pollen and nectar are freely available and there is plenty of perching space, such as on goldenrod (*Solidago* spp.) or cupplant (*Silphium perfoliatum*). With these general guidelines in mind, if there are particular predators or parasitoids that you wish to support, identifying and planting their potential preferred plants may increase their abundance.

CONSIDERATIONS FOR THE HOME GARDEN

As with farms, designing yards and gardens to support a diversity of beneficial insects doesn't mean giving up on other goals you may have: your yard can be beautiful and your garden bountiful while also serving as functional habitat. Here are some design considerations for gardeners hoping to encourage beneficial insects:

- Aim to plant at least 25 percent of your yard with wildflowers, shrubs, or trees that provide stable sources of pollen or nectar.
- Native plants are often the best choice, since they are more likely to promote overall biodiversity and offer advantages such as requiring less water, fertilizer, and pesticides.
- If you prefer nonnative plants for your situation, choose with caution, as many ornamental plants have been bred for showiness, and some varieties produce less pollen or nectar than others. In general, heirloom varieties are preferable to modern varieties.
- Include multiple plant species that bloom throughout the growing season. Early-blooming shrubs like willows or wild lilac and late-blooming wildflowers like asters can provide pollen and nectar at critical times for beneficial insects.
- Intersperse insectary plants with fruit and vegetable plants.
- Ground covers, mulch, leaf litter, or straw can provide shelter for ground-dwelling beneficial insects.
- Provide shelter in the form of a beetle bank, tunnel nest, or insect hotel (see chapter 10).

In most instances, however, the simplest approach is to try to support a wide range of beneficial insects by having several plant species flowering at once, and a sequence of plants flowering through the growing season. If this is your goal, simply select plants with different flower sizes, shapes, and colors, as well as varying plant heights and growth habits.

A LAST THOUGHT ON SELECTING FOR DIVERSITY: Diverse plantings that resemble naturally occurring native plant communities are the most likely to resist pest, disease, and weed epidemics and thus thrive with minimal care. Look around at natural areas in your community to identify which plants are growing together, and which seem to have the most flower visitors.

In the next few chapters, we highlight specific strategies that have been successful for many farmers who are working to enhance beneficial habitat on their land. Those strategies employ the design features we've outlined here, such as flower selection, and the need for insect shelter, to integrate beneficial insect habitat into the broader farm landscape.

Keep in mind that it may take several growing seasons for your habitat improvements to begin to support large numbers of beneficial insects. Some beneficial insects have limited dispersal capacity and reproduction rates, and exhibit slow population growth. For example, although ground beetles are effective predators of aphids, small caterpillars, and other crop pests, they typically only have one generation per growing season.

Goldenrods are among the most common late-blooming wildflowers in much of North America, supporting some of the most diverse groups of beneficial insects.

Exceptions: Native Plants That Host Crop Pests

Primary Crop	Crop Pest or Disease	Known Alternate Hosts of Crop Pests or Diseases
Apples, Pears	Apple Maggot *(Rhagoletis pomonella)*	Hawthorn *(Craetagus* spp.), Wild Plum (*Prunus* spp.)
Apples, Pears	Fire Blight *(Erwinia amylovora)*	Mountain Ash (*Sorbus* spp.), Spirea (*Spiraea* spp.), Hawthorn (*Crataegus* spp.), Cotoneaster (*Cotoneaster* spp.), Toyon (*Heteromeles arbutifolia*), Ocean Spray (*Holodiscus discolor*)
Apples, Pears, Cherries, Peach	Leafroller Caterpillars (several species)	Wild Rose (*Rosa* spp.)
Grapes	Leafhoppers/Sharpshooters (several species)	Willow (*Salix* spp.), Elderberry (*Sambucus* spp.)
Berries	Spotted-Wing Drosophila (*Drosophila suzukii*)	Wild Plum (*Prunus* spp.), Elderberry (*Sambucus* spp.), Wild Raspberry (*Rubus* spp.)

5 Native Plant Field Borders

- Native plant field borders surround the farm with beneficial biodiversity.
- Well-managed field borders can reduce space available for weedy plants.
- Creating a successful native plant field border requires planning and time.
- Once established, field borders are low-maintenance but not no-maintenance.

ON MANY FARMS, border areas along crop fields are an obvious place to establish native grasses and wildflowers for beneficial insects. The border areas' proximity to crops allows easy movement into fields by beneficial insects, especially smaller species that may not travel long distances in search of food.

Native field border plantings can also reduce weed encroachment into crops by occupying ground that might otherwise be dominated by noxious weed species. Most native perennial grasses and wildflowers are slow to establish, and unlikely to become weeds in crop fields. However, where weed control is a concern, maintaining mowed strips between wildflowers and the crop can help reduce the spread of plants into crop fields. These mowed strips can also provide areas for turning equipment without damaging the nearby wildflowers.

An additional benefit of native grass and wildflower field border plantings is that they may reduce the stunting effect that sometimes occurs along the edges of row crops. At least one study identified improved crop yields in edge rows where field border plantings of native grasses and wildflowers were present. Specifically, such field borders reduce crop competition for light, water, and nutrients when compared to field edges occupied by large trees. Additionally, where no trees are present, grass and wildflower field borders can provide a buffer against wind and reduce runoff from crop fields.

This native wildflower field border on a New Hampshire farm required a full season of site preparation and weed control prior to seeding. When the farmer takes the time to deplete the weed seedbank, native wildflowers can grow with little competition.

Field border plantings can take the form of strips along field edges or farm roads, underneath power lines, and can even be established in the corner areas of center-pivot irrigated fields. Because of the location of these areas, and their proximity to crops, it is especially important to protect all field border plantings from pesticide drift, a topic discussed in chapter 11.

Establishment of native grass and wildflower field borders usually requires a full year of site preparation to remove existing weeds and create a clean, relatively weed-free planting site. Site-specific conditions such as soil type, sun or shade exposure, drainage, and local native plant distribution should be factored into field border planting plans. Support for project design and implementation is available from the Xerces Society and the USDA Natural Resources Conservation Service (NRCS). Field border plantings are also a formal NRCS Conservation Practice Standard, making their installation eligible for financial assistance through NRCS-administered conservation programs. Contact your local NRCS office for more information.

Establishing Borders from Seed

Creating wildflower meadows from seed is often the most cost-effective and least arduous approach to creating large-scale beneficial insect habitat. Reseeding a previously weedy field border area with native grasses and wildflowers requires excellent site preparation to reduce preplanting weed pressure, since weed control options are more limited when your wildflowers start to germinate.

Site Preparation

Site preparation is the most important step in creating a native plant field border. It is also a process that may require more than one season of effort to eliminate invasive, noxious, or undesirable nonnative plants prior to seeding. The more time and effort you spend eradicating undesirable plants prior to planting, the more successful your planting is likely to be.

In the instructions below, we describe two very different methods of eliminating weeds prior to planting. Note that we do not recommend tillage in most cases as a site preparation method because tillage tends to stir up dormant weed seed, degrade soil structure, increase erosion, and may harm beneficial soil fungi and other organisms. However, tillage may be appropriate as a site preparation method on land that was previously cropped, grazed, or mowed aggressively for several years, where little weed seed has been added to the upper layer of soil. In all other cases, we recommend one of the following methods.

As an alternative to herbicides, existing vegetation can be killed by solarizing the soil with UV-stabilized greenhouse plastic prior to reseeding an area with wildflowers.

Herbicides are a common way to remove weedy vegetation prior to reseeding an area with wildflowers. The area in this photo was treated several times to kill the existing vegetation. Once replanted, the area will require only occasional spot treatments or hand weeding for maintenance.

SITE PREP: METHOD 1

NONSELECTIVE HERBICIDE

Conventional (nonorganic) farmers may prefer a nonselective herbicide to kill existing weed cover in preparation for reseeding an area. It should only be used on land that has a low risk of erosion, and in areas that are accessible to sprayer equipment.

TIMING

Anticipate spending at least 6 months to reduce weed pressure on typical crop field borders. Begin the first of multiple weed treatments in early spring, after the first weed growth, and plan to seed native plants in the fall.

KEY STEPS

1. Mow any existing thatch as needed before beginning herbicide treatments to expose new weed growth to the herbicide spray.
2. Apply a nonselective herbicide as soon as weeds are actively growing in the early spring.
3. Repeat herbicide applications throughout the spring, summer, and early fall, as needed, or whenever emerging weed seedlings reach 4 to 6 inches (10 to 15 cm).
4. For any herbicide-resistant weeds, mow the area to prevent flowering and seed development as necessary.
5. Plant your native grass/wildflower seed mix in the fall, waiting at least 72 hours after the last herbicide treatment.

NOTE: *Do not till the soil. Avoid any ground disturbance that may bring up dormant weed seed that is buried deeper in the soil. An additional year of site preparation is recommended if weed pressure is particularly high.*

SITE PREP: METHOD 2

SOLARIZATION

Solarization is a great weed removal option for organic farmers and conventional farmers who do not want to use herbicides. It should be used only in areas that (1) have a low risk of erosion, (2) are accessible to mowing equipment, and (3) receive full sun throughout the day.

1.

TIMING

Anticipate around 6 months for the entire process to reduce weed pressure in preparation for seeding. For most parts of North America, begin in the spring, and plan on seeding in the fall.

KEY STEPS

2.

1. Mow and rake the site in the spring, removing any thick stems or branches that might tear the solarization plastic. Then lay UV-stabilized plastic (such as high-tunnel greenhouse plastic), burying the edges to prevent airflow between the plastic and the ground. Weigh down the center of the plastic if necessary to prevent the wind from lifting it. Use greenhouse repair tape for any rips that occur during the season.
2. Remove the plastic in late summer, just before the weather cools, and hand-broadcast wildflower seed onto the soil surface (do not cultivate the soil again after removing the plastic).
3. Mow the seeded area occasionally, if necessary, to prevent fast-growing annual weeds from producing more seed. This will not harm slow-growing perennial wildflowers. Note: If your seed mix consisted mostly of annual wildflowers, skip this step.
4. The field border is dominated by native grasses and wildflowers. Spot-treat or hand-weed as necessary, but avoid disturbing the area.

3.

NOTE: *Do not till the soil. Avoid any major ground disturbance that may bring up additional deeply buried weed seed. Solarization may not be as effective in years when summer sun is limited. An additional year of site preparation is recommended if weed pressure is particularly high.*

4.

Seeding

Below we have outlined several common native grass and wildflower seeding methods. When selecting the method right for you, consider equipment availability and the size of the area you are planting.

PROS AND CONS OF DIFFERENT SEEDING METHODS

Here's an overview of the three seeding methods discussed to help you select the best one for your situation.

Method 1: **Hand-Broadcasting Seed**

PROS

- Hand-broadcasting seed is simple to do and does not require any specialized or expensive equipment.
- Hand broadcasting is an easy way to work with seed that is poorly cleaned or consists of diverse sizes.

CONS

- Hand broadcasting requires a smooth, clean, thatch-free seedbed.
- The seed will benefit from being pressed into the soil after planting for the best success.

Method 2: **Mechanical Drop Seeders or Broadcast Spreaders**

PROS

- Mechanical drop seeders and broadcast seed spreaders come in a variety of sizes, styles, and costs, from low-cost "belly grinder" seeders to walk-behind lawn drop seeders, to tractor or ATV-mounted spreaders that can seed large areas quickly.
- All of these mechanical seeders can provide even seed dispersal over the planting area, and they can usually accommodate both large and small seed, although they may need to be planted in separate batches.

CONS

- Planting with a mechanical seeder requires a smooth, clean, level seedbed.
- Calibration of these seeders can be challenging, and often requires trial and error.
- The seed will also benefit from being pressed into the soil after planting.

Method 3: **No-Till Native Seed Drills**

PROS

- Specialized no-till native seed drills are designed for quick and convenient planting of large areas. Such drills have multiple seed boxes, seed box agitators, and depth controls that are designed specifically for planting small and fluffy native seeds at an optimal rate and depth.
- The major advantage of native seed drills is that you can plant them directly into a light thatch layer, such as vegetation that was killed with herbicides.
- Because seeds are planted in even rows with a native seed drill, it is easy to recognize seedling growth and evaluate establishment success.
- Unlike other planting techniques, native seed drills do not typically benefit from the seed being pressed into the soil surface after planting.

CONS

- No-till native seed drills are expensive and not readily available everywhere, but may sometimes be rented from local conservation districts or hunting groups.
- Specialized native seed drills are also difficult to calibrate for areas less than 1 acre, and require a tractor and an experienced operator. Unless the drill is carefully set up, the seed can easily be planted too deeply, resulting in project failure.

HAND-BROADCASTING SEED

Mix the seed mix and the bulking material to create a homogenous blend.

Because wildflower seed is diverse in size and shape, it is important to mix it adequately.

Remove as much thatch and dead vegetation as possible prior to seeding to create a smooth seedbed. The soil surface can be lightly hand-raked or harrowed to break up crusted surfaces, but avoid cultivation, as it may stir up dormant, buried weed seed.

KEY STEPS

1. Mix your wildflower and grass seed together in a large bucket and bulk up the seed mix with an equal or greater amount of some inert material, such as sand, fine-grained vermiculite, or fine cornmeal. These inert materials ensure that the various types of seed remain evenly distributed in the mix. Bulking up the seed mix with an inert material will also help you see where the mix is being broadcast.
2. Hand broadcast the seed by loosely scattering it on the ground as you might scatter poultry feed. First sow the entire planting area with half of the mix by walking in parallel paths back and forth over the entire site.
3. Next sow the remaining mixture by walking over the entire site again in paths perpendicular to the first sowing. This will help ensure that seed is evenly distributed over the entire area.
4. After scattering the seed, do not cover it with soil. Instead, if possible, use a water-filled turf grass roller (available for rent at most hardware stores) or a cultipacker to press the seed into the soil surface. Natural precipitation or light overhead irrigation can also help ensure good seed-to-soil contact.
5. Use floating row-cover fabric if necessary to protect seeds and small seedlings against bird predation.

Add fine sand, cornmeal, or vermiculite to the seed mix to increase its volume and make hand broadcasting easier.

Divide the seed mix blend into two different batches.

Hand-scatter the seed like chicken feed directly on the soil surface. Broadcast one batch first, then the second while walking in a perpendicular direction to your first seeding path.

If possible, it is useful to press the seed into the soil surface with a cultipacker or lawn roller to ensure good seed-to-soil contact.

SEEDING: METHOD 2

MECHANICAL DROP SEEDERS OR BROADCAST SPREADERS

Remove as much thatch and dead vegetation as possible prior to seeding to create a smooth seedbed. The soil surface can be lightly hand-raked or harrowed to break up crusted surfaces, but avoid cultivation, as it may stir up dormant, buried weed seed.

KEY STEPS

Seed of similar sizes can be mixed together and bulked up with an inert carrier ingredient such as sand, fine-grained vermiculite, or fine cornmeal for even seed distribution and ease of calibration.

1. Begin planting with the flow gate set to the narrowest opening, to allow multiple passes over the seedbed for even distribution.
2. Plant very large seed separately, with the flow gate set to a wider opening.
3. After scattering the seed, do not cover it with soil. Instead, if possible, use a water-filled turf grass roller (available for rent at most hardware stores) or a cultipacker to press the seed into the soil surface. Natural precipitation or light overhead irrigation can also help ensure good seed-to-soil contact.
4. Use floating row-cover fabric if necessary to protect seeds and small seedlings against bird predation.

Mix the seed with fine sand, cornmeal, or vermiculite to increase its volume and make broadcasting easier.

Weed Control during Seedling Establishment

Weed control is critical in the first few years after planting. If your site was well prepared from the beginning, then you'll need to expend less effort for weeding after your native field border is installed. It isn't reasonable to expect your native field border to be weed-free, even in the long term, but if you can prevent most weeds from going to seed in, or adjacent to, the project area during the first two years after planting, you can usually anticipate long-term success.

Familiarity with the life cycles of your common weeds will help you time weed control practices, and since young wildflower and weed seedlings may look alike, take care to properly identify weeds before you remove them.

SEEDING: METHOD 3

NO-TILL NATIVE SEED DRILLS

With a native seed drill, you should only plant when the soil is dry enough to prevent sticking to the coulters. Under wet conditions, small seed is likely to stick to mud-caked parts of the drill rather than the ground.

KEY STEPS

1. Loosely fill the seed boxes, but do not compact seed into them. Seed quantities that do not cover the agitator should be planted using some other method, since the drill is difficult to calibrate for small volumes of seed.
2. For most wildflower species, set the depth controls to plant no more than ¼ inch (6 mm) deep (consult the seed vendor for specific guidelines on very sandy soils), and stop periodically to check the planting depth. Some seed should be observable on or just below the soil surface. As a general rule, planting depth should be equal to 1.5x the diameter of the seed.
3. Operate the drill at less than 5 mph, checking periodically for clogged planting tubes (usually observed as a seed box that remains full). Clogging is common with fluffy seed, or seed with a lot of chaff. Also, avoid backing up, as doing so can cause clogging.
4. Consult the owner's manual for information on seed drill calibration and operation.

To limit soil disturbance in an area with existing vegetation, consider using a no-till native seed drill.

Common weed-management strategies include:

SPOT SPRAYING. Using herbicides to target specific areas can be effective, relatively inexpensive, and requires minimal labor, even on larger project areas. Be careful to prevent herbicides from dripping onto desirable plant species. Spot spraying is usually performed with a backpack sprayer.

MOWING/STRING TRIMMING. To keep weedy species from shading out other plants, and to prevent them from going to seed, you can mow or string-trim your field borders.

Mowing is especially useful when you're establishing perennial wildflower plots. After planting a site with perennial seed mix, mow it regularly, ideally at 6 inches (15 cm) or higher,

Mowing habitat in the first year or two after planting is a common weed-management strategy. Here, invasive yellow star thistle is being mowed to allow the recently planted, slower-growing native wildflowers to capture more sunlight.

Inter-seeding additional wildflower seed into existing habitat can increase plant diversity. Here a heavy disk is used to open up spaces in the existing grass cover for additional wildflower seed to be broadcast.

during the first year after planting to prevent annual and biennial weeds from flowering and producing seed. Perennial wildflowers are slow to establish from seed, and are usually not harmed by incidental mowing in the first year after planting. You can also mow plots of reseeding annuals at the end of the growing season to help spread wildflower seeds, and to prevent tree seedlings from establishing themselves in your meadow.

HAND WEEDING. Hand weeding, including hoeing, can be effective in small areas with moderate to low weed pressure.

Long-Term Maintenance

As your field border meadow continues to mature, be sure to protect it from insecticide and herbicide drift, unless you need herbicides to control invasive plants. Occasional hand weeding may also be necessary to control noxious weeds. If plant diversity declines after several years, you can often restore it by interplanting additional wildflower seed into strips that have been closely mowed or disked (to less than 1 inch (2.5 cm) deep).

Native plant field borders also sometimes need management to prevent weedy tree seedlings from establishing and taking over. Both mowing and burning are common strategies for removing tree seedlings and renewing some of the original plant vigor. If you mow, be sure all your equipment is clean and free of weed seed, and do not mow or burn during critical wildlife nesting seasons, usually in the spring and summer. After establishment, no more than 30 percent of the field border area should be mowed or burned in any one year to ensure sufficient undisturbed food and shelter for your beneficial insects. (See chapter 12 for more information.)

Sample Seed Mixes for Native Plant Field Borders

The seed mixes in this section represent examples of what you might consider planting to provide beneficial insect habitat in your own region. The species included in these sample mixes are usually available from commercial producers, and have been used by Xerces Society conservation technicians in recent years. We hope that this gives you a sense of what to anticipate in terms of the quantities you will need.

A few additional things that you should know:

- Many more details about each of these species are available in part 5 of this book.
- All of these sample mixes are formulated for a 1-acre planting area.
- Species marked with * are sometimes easier to establish from transplants than from seed.
- Some older native grass and wildflower seeding recommendations call for seeding rates of a specific weight of seed per acre. We find that calculating rates in terms of seeds per square foot gives more satisfactory results. Based on our field trials nationwide, rates of 40 to 80 seeds per square foot are consistently the best. The rates here are formulated at 60 seeds per square foot.
- Although the species in these sample mixes are all native to their respective region, they may not be native to your specific location. If this is a concern (e.g., if you are located near sensitive natural areas), please omit plants that are not locally native.
- Many other species not included in these mixes may also be appropriate in a beneficial insect seed mix. Explore what is available to you, and consider adding more diversity whenever you can.
- While native seed costs may seem expensive, remember that the costs of establishing native plant field borders (including the cost of seed) often can be partially offset by financial assistance programs available through the USDA Natural Resources Conservation Service (NRCS). Consult your local NRCS Service Center to find out what is available to you.

NATIVE BORDERS FOR THE HOME GARDEN

Just like larger farm fields, even small gardens can benefit from native plant borders. Aim for a diversity of wildflowers that provide pollen and nectar throughout the growing season, as well as a few clump-forming grasses. Those plants, along with strategies such as cover-cropping and providing shelter (see chapters 8 and 10), can ensure all the pest control many gardeners need.

Northeastern United States

Common Name	Scientific Name	Percent of Mix	Seeds per Square Foot
NATIVE WILDFLOWERS			
Canada Anemone*	*Anemone canadensis*	1 percent	0.6
Golden Alexanders	*Zizia aurea*	2 percent	1.2
Lanceleaf Coreopsis	*Coreopsis lanceolata*	10 percent	6
Yarrow	*Achillea millefolium*	5 percent	3
Partridge Pea	*Chamaecrista fasciculata*	6 percent	3.6
Butterfly Milkweed*	*Asclepias tuberosa*	1 percent	0.6
Virginia Mountain Mint	*Pycnanthemum virginianum*	10 percent	6
Dotted Mint	*Monarda punctata*	25 percent	15
Boneset	*Eupatorium perfoliatum*	5 percent	3
Showy Goldenrod	*Solidago speciosa*	5 percent	3
Calico Aster	*Symphyotrichum lateriflorum*	10 percent	6
NATIVE GRASSES			
Little Bluestem	*Schizachyrium scoparium*	10 percent	6
Big Bluestem	*Andropogon gerardii*	5 percent	3
Indian Grass	*Sorghastrum nutans*	5 percent	3
TOTALS		**100 percent**	**60 Seeds per Square Foot**

**These species are sometimes easier to establish from transplants than from seed.*

Canada Anemone

Big Bluestem

Butterfly Milkweed

	Seeds per Pound	Pounds Needed	Bloom Period	Life Cycle
	128,000	0.20	Spring	Perennial
	150,000	0.35	Spring	Perennial
	320,000	0.82	Early Summer	Perennial
	2,000,000	0.07	Summer	Perennial
	60,000	2.61	Summer	Annual
	70,000	0.37	Summer	Perennial
	3,850,000	0.07	Summer	Perennial
	1,450,000	0.45	Summer	Perennial
	2,000,000	0.07	Summer	Perennial
	1,300,000	0.10	Autumn	Perennial
	2,260,000	0.12	Autumn	Perennial
	200,000	1.31	—	Perennial
	140,000	0.93	—	Perennial
	175,000	0.75	—	Perennial
		8.22 Pounds per Acre		

Mountain Mint

Partridge Pea

Indian Grass

Southeastern United States

Common Name	Scientific Name	Percent of Mix	Seeds per Square Foot
NATIVE WILDFLOWERS			
Lanceleaf Coreopsis	*Coreopsis lanceolata*	6 percent	3.6
Butterfly Milkweed*	*Asclepias tuberosa*	0.5 percent	0.3
Partridge Pea	*Chamaecrista fasciculata*	4 percent	2.4
Dotted Mint	*Monarda punctata*	15 percent	9
Annual Blanketflower	*Gaillardia pulchella*	12 percent	7.2
Cupplant*	*Silphium perfoliatum*	0.5 percent	0.3
Virginia Mountain Mint	*Pycnanthemum virginianum*	12 percent	7.2
Rattlesnake Master	*Eryngium yuccifolium*	3 percent	1.8
Swamp Sunflower	*Helianthus angustifolius*	12 percent	7.2
Pine Barren Goldenrod	*Solidago fistulosa*	7 percent	4.2
Anise Scented Goldenrod	*Solidago odora*	6 percent	3.6
NATIVE GRASSES			
Little Bluestem	*Schizachyrium scoparium*	10 percent	6
Eastern Gamagrass	*Tripsacum dactyloides*	1 percent	0.6
Wiregrass	*Aristida stricta*	3 percent	1.8
Indian Grass	*Sorghastrum nutans*	8 percent	4.8
TOTALS		**100 percent**	**60 Seeds per Square Foot**

**These species are sometimes easier to establish from transplants than from seed.*

Wiregrass

Cupplant

Rattlesnake Master

	Seeds per Pound	Pounds Needed	Bloom Period	Life Cycle
	221,000	2.13	Early Summer	Perennial
	70,000	0.37	Summer	Perennial
	60,000	2.18	Summer	Annual
	1,440,000	0.36	Summer	Perennial
	230,000	1.14	Summer	Annual
	33,600	0.39	Summer	Perennial
	3,850,000	0.07	Summer	Perennial
	127,680	0.41	Late Summer	Perennial
	504,000	0.13	Late Summer	Perennial
	700,000	0.22	Autumn	Perennial
	1,000,000	0.13	Autumn	Perennial
	200,000	1.31	—	Perennial
	7,000	3.73	—	Perennial
	750,000	0.10	—	Perennial
	175,000	1.19	—	Perennial
		13.86 Pounds per Acre		

Dotted Mint

Annual Blanketflower

Eastern Gamagrass

Midwest

Common Name	Scientific Name	Percent of Mix	Seeds per Square Foot
NATIVE WILDFLOWERS			
Golden Alexanders	*Zizia aurea*	1 percent	0.6
Lanceleaf Coreopsis	*Coreopsis lanceolata*	10 percent	6
Dotted Mint	*Monarda punctata*	15 percent	9
Butterfly Milkweed*	*Asclepias tuberosa*	0.5 percent	0.3
Common Milkweed*	*Asclepias syriaca*	0.5 percent	0.3
Yarrow	*Achillea millefolium*	10 percent	6
Partridge Pea	*Chamaecrista fasciculata*	5 percent	3
Cupplant*	*Silphium perfoliatum*	0.5 percent	0.3
Virginia Mountain Mint	*Pycnanthemum virginianum*	15 percent	9
Rattlesnake Master	*Eryngium yuccifolium*	0.5 percent	0.3
Maximilian Sunflower	*Helianthus maximiliani*	5 percent	3
Showy Goldenrod	*Solidago speciosa*	2 percent	1.2
Calico Aster	*Symphyotrichum lateriflorum*	5 percent	3
NATIVE GRASSES			
Big Bluestem	*Andropogon gerardii*	5 percent	3
Prairie Junegrass	*Koeleria macrantha*	12 percent	7.2
Little Bluestem	*Schizachyrium scoparium*	12 percent	7.2
Prairie Dropseed	*Sporobolus heterolepis*	1 percent	0.6
TOTALS		**100 percent**	**60 Seeds per Square Foot**

**These species are sometimes easier to establish from transplants than from seed.*

Showy Goldenrod

Prairie Junegrass

Little Bluestem

Seeds per Pound	Pounds Needed	Bloom Period	Life Cycle
200,000	0.13	Spring	Perennial
320,000	0.82	Early Summer	Perennial
1,400,000	0.27	Summer	Perennial
55,680	0.23	Summer	Perennial
64,000	0.20	Summer	Perennial
2,000,000	0.13	Summer	Perennial
60,000	2.18	Summer	Annual
33,600	0.39	Summer	Perennial
5,300,000	0.07	Summer	Perennial
127,680	0.10	Late Summer	Perennial
180,000	0.73	Late Summer	Perennial
1,680,000	0.03	Autumn	Perennial
2,260,000	0.06	Autumn	Perennial
140,000	0.93	—	Perennial
2,400,000	0.13	—	Perennial
200,000	1.57	—	Perennial
256,000	0.10	—	Perennial
	8.07 Pounds per Acre		

Prairie Dropseed

Common Milkweed

Lanceleaf Coreopsis

Northern Plains

Common Name	Scientific Name	Percent of Mix	Seeds per Square Foot
NATIVE WILDFLOWERS			
Yarrow	*Achillea millefolium*	15 percent	9
Perennial Blanketflower	*Gaillardia aristata*	12 percent	7.2
Showy Milkweed	*Asclepias speciosa*	1 percent	0.6
Maximilian Sunflower	*Helianthus maximiliani*	10 percent	6
Common Sunflower	*Helianthus annuus*	8 percent	4.8
Plains Coreopsis	*Coreopsis tinctoria*	15 percent	9
Oldfield Goldenrod	*Solidago nemoralis*	6 percent	3.6
Smooth Blue Aster	*Symphyotrichum laeve*	8 percent	4.8
NATIVE GRASSES			
Prairie Junegrass	*Koeleria macrantha*	4 percent	2.4
Little Bluestem	*Schizachyrium scoparium*	8 percent	4.8
Big Bluestem	*Andropogon gerardii*	5 percent	3
Slender Wheatgrass	*Elymus trachycaulus*	8 percent	4.8
TOTALS		**100 percent**	**60 Seeds per Square Foot**

Yarrow

Maximilian Sunflower

Showy Milkweed

	Seeds per Pound	Pounds Needed	Bloom Period	Life Cycle
	2,000,000	0.20	Summer	Perennial
	132,000	2.38	Summer	Perennial
	72,000	0.36	Summer	Perennial
	180,000	1.45	Summer	Perennial
	59,800	3.5	Late Summer	Annual
	1,400,000	0.28	Late Summer	Annual
	300,000	0.52	Autumn	Perennial
	880,000	0.24	Autumn	Perennial
	2,400,000	0.04	—	Perennial
	200,000	1.05	—	Perennial
	140,000	0.93	—	Perennial
	135,000	1.55	—	Perennial
		12.5 Pounds per Acre		

Slender Wheatgrass

Plains Coreopsis

Smooth Blue Aster

Southern Plains

Common Name	Scientific Name	Percent of Mix	Seeds per Square Foot
NATIVE WILDFLOWERS			
Butterfly Milkweed*	*Asclepias tuberosa*	0.5 percent	0.3
Showy Milkweed*	*Asclepias speciosa*	1 percent	0.6
Yarrow	*Achillea millefolium*	12 percent	7.2
Perennial Blanketflower	*Gaillardia aristata*	15 percent	9
Partridge Pea	*Chamaecrista fasciculata*	6 percent	3.6
Lemon Beebalm	*Monarda citriodora*	12.5 percent	7.5
Maximilian Sunflower	*Helianthus maximiliani*	11 percent	6.6
Plains Coreopsis	*Coreopsis tinctoria*	10 percent	6
Common Sunflower	*Helianthus annuus*	5 percent	3
Oldfield Goldenrod	*Solidago nemoralis*	4 percent	2.4
NATIVE GRASSES			
Side Oats Grama	*Bouteloua curtipendula*	7 percent	4.2
Big Bluestem	*Andropogon gerardii*	6 percent	3.6
Little Bluestem	*Schizachyrium scoparium*	10 percent	6
TOTALS		**100 percent**	**60 Seeds per Square Foot**

Intermountain West

Common Name	Scientific Name	Percent of Mix	Seeds per Square Foot
Yarrow	*Achillea millefollium*	10 percent	6
Showy Milkweed*	*Asclepias speciosa*	2 percent	1.2
Perennial Blanketflower	*Gaillardia aristata*	20 percent	12
Sulfur Buckwheat*	*Eriogonum umbellatum*	10 percent	6
Plains Coreopsis	*Coreopsis tinctoria*	6 percent	3.6
Common Sunflower	*Helianthus annuus*	5 percent	3
Canada Goldenrod	*Solidago canadensis*	10 percent	6
Smooth Blue Aster	*Symphyotrichum laeve*	11 percent	6.6
Idaho Fescue	*Festuca idahoensis*	10 percent	6
Blue Bunch Wheatgrass	*Pseudoroegneria spicata*	8 percent	4.8
Blue Wild Rye	*Elymus glaucus*	8 percent	4.8
TOTALS		**100 percent**	**60 Seeds per Square Foot**

**These species are sometimes easier to establish from transplants than from seed.*

	Seeds per Pound	Pounds Needed	Bloom Period	Life Cycle
	55,680	0.23	Summer	Perennial
	72,000	0.36	Summer	Perennial
	2,000,000	0.16	Summer	Perennial
	132,000	2.97	Summer	Perennial
	60,000	2.61	Summer	Annual
	819,000	0.4	Summer	Annual
	180,000	1.60	Late Summer	Perennial
	1,400,000	0.19	Late Summer	Annual
	59,800	2.19	Late Summer	Annual
	300,000	0.35	Autumn	Perennial
	640,000	0.29	—	Perennial
	140,000	1.12	—	Perennial
	200,000	1.31	—	Perennial
		13.78 Pounds per Acre		

Lemon Beebalm

Sulfur Buckwheat

	Seeds per Pound	Pounds Needed	Bloom Period	Life Cycle
	2,770,000	0.09	Summer	Perennial
	72,000	0.73	Summer	Perennial
	132,000	3.96	Summer	Perennial
	210,000	1.24	Summer	Perennial
	1,400,000	0.11	Late Summer	Annual
	59,800	2.19	Late Summer	Annual
	4,600,000	0.06	Autumn	Perennial
	880,000	0.33	Autumn	Perennial
	450,000	0.58	—	Perennial
	140,000	1.49	—	Perennial
	120,000	1.74	—	Perennial
		12.52 Pounds per Acre		

Canada Goldenrod

Southwestern United States

Common Name	Scientific Name	Percent of Mix	Seeds per Square Foot
Globe Gilia	*Gilia capitata*	4 percent	2.4
California Bluebells	*Phacelia campanularia*	5 percent	3
Lacy Phacelia	*Phacelia tanacetifolia*	15 percent	9
Yarrow	*Achillea millefolium*	8 percent	4.8
Showy Milkweed*	*Asclepias speciosa*	5 percent	3
Lemon Beebalm	*Monarda citriodora*	6 percent	3.6
Perennial Blanketflower	*Gaillardia aristata*	10 percent	6
Sulfur Buckwheat*	*Eriogonum umbellatum*	6 percent	3.6
Common Sunflower	*Helianthus annuus*	6 percent	3.6
Plains Coreopsis	*Coreopsis tinctoria*	5 percent	3
Canada Goldenrod	*Solidago canadensis*	6 percent	3.6
Idaho Fescue	*Festuca idahoensis*	16 percent	8.8
Slender Wheatgrass	*Elymus trachycaulus*	8 percent	4.8
TOTALS		**100 percent**	**60 Seeds per Square Foot**

California

Common Name	Scientific Name	Percent of Mix	Seeds per Square Foot
Douglas Meadowfoam	*Limnanthes douglasii*	3 percent	1.8
California Bluebells	*Phacelia campanularia*	6 percent	3.6
Globe Gilia	*Gilia capitata*	8 percent	4.8
California Phacelia	*Phacelia californica*	10 percent	6
Lacy Phacelia	*Phacelia tanacetifolia*	8 percent	4.8
Yarrow	*Achillea millefolium*	5 percent	3
Narrowleaf Milkweed	*Asclepias fascicularis*	1 percent	0.6
Sulfur Buckwheat*	*Eriogonum umbellatum*	5 percent	3
Showy Milkweed*	*Asclepias speciosa*	2 percent	1.2
Common Sunflower	*Helianthus annuus*	9 percent	0.3
Bolander's Sunflower	*Helianthus bolanderi*	1 percent	0.6
Canada Goldenrod	*Solidago canadensis*	8 percent	4.8
Pacific Aster*	*Symphyotrichum chilense*	10 percent	6
Roemer's Fescue	*Festuca roemeri*	9 percent	5.4
Idaho Fescue	*Festuca idahoensis*	9 percent	5.4
California Oatgrass	*Danthonia californica*	6 percent	3.6
TOTALS		**100 percent**	**60 Seeds per Square Foot**

These species are sometimes easier to establish from transplants than from seed.

Seeds per Pound	Pounds Needed	Bloom Period	Life Cycle
800,000	0.13	Spring	Annual
750,000	0.17	Spring	Annual
223,300	1.76	Spring	Annual
2,770,000	0.08	Summer	Perennial
72,000	1.82	Summer	Perennial
819,000	0.19	Summer	Annual
132,000	1.98	Summer	Annual
210,000	0.75	Summer	Perennial
59,800	2.62	Late Summer	Annual
1,400,000	0.09	Late Summer	Annual
4,600,000	0.03	Autumn	Perennial
450,000	0.92	—	Perennial
135,000	1.55	—	Perennial
	12.48 Pounds per Acre		

Roemer's Fescue

Lacy Phacelia

Seeds per Pound	Pounds Needed	Bloom Period	Life Cycle
86,400	0.91	Spring	Annual
750,000	0.21	Spring	Annual
800,000	0.26	Spring	Annual
250,800	0.52	Spring	Perennial
223,300	0.94	Spring	Annual
2,770,000	0.05	Summer	Perennial
64,000	0.41	Summer	Perennial
210,000	0.62	Summer	Perennial
72,000	0.73	Summer	Perennial
59,800	0.22	Late Summer	Annual
104,000	0.25	Late Summer	Annual
4,600,000	0.05	Autumn	Perennial
2,668,000	0.10	Autumn	Perennial
500,000	0.47	—	Perennial
450,000	0.52	—	Perennial
120,000	1.31	—	Perennial
	7.57 Pounds per Acre		

Common Sunflower

Showy Milkweed

Pacific Northwest

Common Name	Scientific Name	Percent of Mix	Seeds per Square Foot
Douglas Meadowfoam	*Limnanthes douglasii*	10 percent	6
Globe Gilia	*Gilia capitata*	9.5 percent	5.7
Self-Heal	*Prunella vulgaris*	15 percent	9
Yarrow	*Achillea millefolium*	5 percent	3
Perennial Blanketflower	*Gaillardia aristata*	10 percent	6
Showy Milkweed*	*Asclepias speciosa*	2 percent	1.2
Sulfur Buckwheat*	*Eriogonum umbellatum*	10 percent	6
Douglas Aster*	*Symphyotrichum subspicatum*	0.5 percent	0.3
Canada Goldenrod	*Solidago canadensis*	12 percent	7.2
Prairie Junegrass	*Koeleria macrantha*	10 percent	6
Roemer's Fescue	*Festuca romeri*	8 percent	4.8
Blue Wild Rye	*Elymus glaucus*	8 percent	4.8
TOTALS		**100 percent**	**60 Seeds per Square Foot**

These species are sometimes easier to establish from transplants than from seed.

Perennial Blanketflower

Douglas Meadowfoam

Globe Gilia

Seeds per Pound	Pounds Needed	Bloom Period	Life Cycle
86,400	3.03	Spring	Annual
800,000	0.31	Spring	Annual
400,000	0.98	Early Summer	Perennial
2,000,000	0.07	Summer	Perennial
132,000	1.98	Summer	Perennial
72,000	0.73	Summer	Perennial
210,000	1.24	Summer	Perennial
1,120,000	0.01	Autumn	Perennial
3,000,000	0.10	Autumn	Perennial
2,315,000	0.11	—	Perennial
500,000	0.42	—	Perennial
120,000	1.74	—	Perennial
	10.72 Pounds per Acre		

Self-Heal

Blue Wild Rye

Yarrow

6 Insectary Strips

- Insectary strips bring beneficial insect habitat into the crop field itself.
- Insectary strips are low cost and easy to create.
- The best location for insectary strips is on farms that don't use insecticides.
- When you are ready to rotate crops, insectary strips can simply be plowed under and replanted elsewhere.

INSECTARY STRIPS are a conservation biocontrol strategy that locates pollen and nectar resources within the crop itself. This strategy is used, for example, by organic lettuce producers in California who plant entire rows of sweet alyssum at regular intervals between rows of lettuce to support aphid predators and parasitoids.

*Plains coreopsis (*Coreopsis tinctoria*) is an annual native wildflower that attracts beneficial insects such as predatory wasps.*

*Native to California and the Southwest, California bluebells (*Phacelia campanularia*) is a fast-growing annual that blooms exceptionally early in warm climates.*

*Herbs like dill (*Anethum graveolens*) are low cost and highly attractive to beneficial insects such as flower flies.*

Because mass insectary plantings are usually temporary, and placed within crop fields that will later be cultivated for a new crop, farmers who use them usually focus on low-cost, rapid-blooming annual species (although perennial species are also an option for permanent installations). Nonnative species that are typically recommended include alyssum (*Lobularia maritima*), bachelor button (*Centaurea cyanus*), buckwheat (*Fagopyrum esculentum*), marigold (*Tagetes* spp.), calendula (*Calendula* spp.), dill (*Anethum graveolens*), cilantro (*Coriander sativum*), and lacy phacelia (*Phacelia tanacetifolia*), a California native used worldwide for insectary or pollinator plantings. Though widely available at affordable prices, we consider marigold and calendula to be much less valuable to beneficial insects than other options, such as buckwheat.

Insectary plantings are usually established at the time of the primary crop planting and plowed up at the end of the season. While they can significantly enhance populations of pollen and nectar feeding beneficial insects, their temporary nature does not offer insects shelter for overwintering. Combined with permanent native grass and wildflower field borders, however, mass insectary plantings may provide corridors that help move beneficial insects out of field edges and directly into the crop itself, enhancing pest control throughout the entire field.

Since insectary plantings are usually integrated into crop fields, it is imperative that they not be exposed to insecticides. For this reason they are usually only used by farmers who do not apply insecticides.

Most insectary strips consist of fast-growing annuals planted within or alongside the primary crop.

Perennials or Annuals?

Although most insectary plantings consist of nonnative annual wildflowers and herbs, they can also be established on a more permanent basis with perennial wildflowers and grasses instead.

The advantage of annuals is that they bloom the first year, and usually don't persist to become weeds later on. Depending on the species, you may get some reseeding, resulting in an expected life span for the strip of a year or two, unless you plow the flowers under or cut them back before they set seed. One drawback of annual flower strips is that they often require more water than native perennial wildflowers.

Perennial wildflowers established from seed take longer to develop, and some species may not produce any flowers for two or three years. The trade-off is that a perennial insectary strip may last for many years. Perennial insectary strips also provide an opportunity to include native bunch grasses for overwintering and insect reproduction, a benefit that is difficult to accommodate with temporary strips of annual wildflowers. Unlike annual wildflowers, perennials usually don't require irrigation after they are established; however, they will require routine weeding.

The deciding factor in selecting a perennial mix over an annual one is often simply whether you intend to plow up the area after a single season and replant it with other crops. In such cases, annual strips are the clear choice. Ultimately, however, you might think about using a combination of both on your farm.

Insectary strips are traditionally temporary features planted with quick-growing annual species that provide pollen and nectar soon after planting.

Mix seed of multiple species together and plant in single rows between crops.

This insectary strip in California provides a huge mass of pollen and nectar for beneficial insects throughout the growing season. When the field is ready for replanting or a crop rotation, the insectary strip can simply be plowed under.

HOW TO

PLANT INSECTARY STRIPS

METHOD

- Plant annual strips in the spring after the danger of frost is over. Plant perennial strips in late fall so that the seed can be cold-stratified for maximum germination.
- Prepare the ground as you would for any other crop.
- Plant insectary seed mixes either by hand scattering, or by using a vegetable seed drill if you want the strip planted as a row between crop plants.
- As a general rule, native perennial wildflowers should be planted at or just below the soil surface. Seed for nonnative annuals should be just barely covered with soil.

Note: Perennial strips should be planted on land that has been cultivated for multiple seasons and is relatively weed-free, since native perennials usually grow more slowly than weeds.

If it is too difficult to plant the seed as a mixture in a narrow space, it is no problem to plant each species separately, side by side, in rows. The insects won't care, although a dense mixture of plants is more likely to suppress weed growth.

Farms that specialize in annual row crops can easily integrate insectary strips throughout their operations, including around and even inside high-tunnel greenhouses.

CASE STUDY

An Insectary Seed Mix for New Mexico Pumpkins

COMMERCIAL INSECTARY seed mixes for conservation biocontrol do not always perform consistently well in different regions. In preliminary trials, for example, three such mixes failed to establish in the alkaline soils and flood irrigation systems typical of much of New Mexico. We developed an insectary mix suitable for use under these conditions by testing various mixtures of eight or nine species of flowering annuals over three years in both clay and sandy soils, using pumpkins as a model cropping system.

Ideally, insectary plantings should consist of a mixture of species that vary in flower structure, size, color, and bloom period, to benefit the maximum number of beneficial insects while not stimulating increases in pest species. For our trials, we focused on annual garden flowers and culinary herbs with inexpensive, readily available seed that would be quick to bloom, flower for a prolonged period, and would be suitable for use as part of an annual crop rotation for small-scale vegetable growers. In addition to developing a suitable insectary mix, we also assessed the effect of the mixtures on populations of various beneficial insects and the two principal pests associated with pumpkins in New Mexico, squash bug (*Anasa tristis*) and spotted cucumber beetle (*Diabrotica undecimpunctata howardi*).

After three years of trials, we propose a core mix of the following six species for New Mexico: California bluebell (*Phacelia campanularia*), buckwheat (*Fagopyrum esculentum*), dill (*Anethum graveolens*), plains coreopsis (*Coreopsis tinctoria*), garden cosmos (*Cosmos bipinnatus*), and alyssum (*Lobularia maritima*). In general, these core species performed well, although on sandy soils buckwheat proved to be highly susceptible to root rots caused by *Rhizoctonia solani*, while on clay soils cosmos was too vigorous, dominating the mix and shading out some of the lower-growing species. In addition, alyssum, a member of the brassica family, attracted pests such as flea beetles, harlequin bug (*Murgantia histrionica*), and the exotic invasive bagrada bug (*Bagrada hilaris*); alyssum should therefore be omitted where brassica crops such as arugula and mustard greens are grown.

Research in New Mexico found that spotted cucumber beetle numbers were lower in pumpkins grown with an insectary mix consisting of six types of flowers.

INSECTARY PLANTS IN THE HOME GARDEN

Adding insectary plants in your home garden can be as simple as planting a single row of flowers in the middle of your vegetables, or as sophisticated as establishing a flowering perimeter that surrounds the entire garden. For quick and easy insectary plants, many of the same herbs you grow for the kitchen, especially dill and cilantro, will attract huge numbers of flower flies and parasitoid wasps if you let them flower.

In our trials, combined populations of all species of parasitic wasps were higher in plots with insectary plantings than in control plots in all three years, while populations of beneficial pirate bugs and adult lacewings showed the same trend in two of the three years, and the convergent lady beetle (*Hippodamia convergens*) in just one of the years. Spotted cucumber beetles were lower in plots with insectary plantings in two of the years, but squash bug numbers were not affected at all. The most common natural enemies of squash bug were two species of squash bug egg parasitoids, which colonized the plots only relatively late in the season. We believe that providing stable overwintering habitat would keep the wasps closer to the crop fields and thus improve the performance of squash bug parasitoids.

— DR. TESSA R. GRASSWITZ, New Mexico State University
Los Lunas Agricultural Science Center

Sample Insectary Seed Mixes

The following sample mixes represent both annual and perennial species mixes for the eastern and western United States. They are formulated to densely cover an area size of 1,000 square feet. All of these species are readily available from mail order and Internet-based bulk wildflower seed companies.

Wildflower seeds ready for planting.

Annual Mix, Eastern United States

ADVANTAGES: This mix matures and blooms quickly and the plants won't become long-term weeds.

DISADVANTAGES: Even if allowed to reseed, this mix will only last one to two years. The plants will also require irrigation in most areas.

Common Name	Scientific Name	Percent of Mix	Seeds per Square Foot	Seeds per Ounce	Ounces Needed
Bachelor Button	*Centaurea cyanus*	10 percent	8	6,000	1.75
Dill	*Anethum graveolens*	10 percent	8	10,000	1.0
Cilantro	*Coriander sativum*	7 percent	5.6	2,000	3.5
Plains Coreopsis	*Coreopsis tinctoria*	25 percent	20	87,500	0.25
Partridge Pea	*Chamaecrista fasciculata*	5 percent	4	3,750	1.5
Annual Blanketflower	*Gaillardia pulchella*	10 percent	8	14,380	1.0
Alyssum*	*Lobularia maritima*	8 percent	6.4	70,880	0.25
Buckwheat	*Fagopyrum esculentum*	10 percent	8	890	12.0
Lemon Beebalm	*Monarda citriodora*	10 percent	8	52,440	0.25
Annual Sunflower	*Helianthus annuus*	5 percent	4	2,810	2.0
TOTALS		**100 percent**	**80**	**250,650**	**23.5**

**May attract pests of brassica crops (e.g., broccoli, arugula, etc.) like flea beetles and harlequin bug. Omit from the mix if this is a concern.*

Perennial Mix, Eastern United States

ADVANTAGES: The expected life span of this mix is at least five years, and after establishment it won't require irrigation in most locations.

DISADVANTAGES: Perennial wildflowers planted from seed are unlikely to bloom for at least a full year. Perennial wildflowers are also more sensitive to proper planting time and seeding depth, so surface planting is often best.

Common Name	Scientific Name	Percent of Mix	Seeds per Square Foot	Seeds per Ounce	Ounces Needed
Dotted Mint	*Monarda punctata*	15 percent	12	90,625	0.25
Yarrow	*Achillea millefolium*	15 percent	12	171,875	0.25
Lanceleaf Coreopsis	*Coreopsis lanceolata*	15 percent	12	13,810	1.25
Perennial Blanketflower	*Gaillardia aristata*	10 percent	8	11,625	1.0
Alfalfa	*Medicago sativa*	15 percent	12	14,190	1.25
Little Bluestem	*Schizachyrium scoparium*	15 percent	12	12,500	1.25
Canada Wild Rye	*Elymus canadensis*	15 percent	12	7,125	1.5
TOTALS		**100 percent**	**80**		**6.75**

Annual Mix, Western United States

ADVANTAGES: This mix matures and blooms quickly and the plants won't become long-term weeds.

DISADVANTAGES: Even if allowed to reseed, this mix will only last one to two years. The plants will also require irrigation in most areas.

Common Name	Scientific Name	Percent of Mix	Seeds per Square Foot	Seeds per Ounce	Ounces Needed
Bachelor Button	*Centaurea cyanus*	10 percent	8	6,000	1.75
Dill	*Anethum graveolens*	16 percent	12.8	10,000	1.5
Cilantro	*Coriander sativum*	10 percent	8	2,000	5.25
Plains Coreopsis	*Coreopsis tinctoria*	10 percent	8	87,500	0.25
Hubam Sweetclover	*Melilotus officinalis 'Hubam'*	8 percent	6.4	16,250	0.5
Annual Blanketflower	*Gaillardia pulchella*	10 percent	8	14,380	1.0
Alyssum*	*Lobularia maritima*	8 percent	6.4	70,880	0.25
Buckwheat	*Fagopyrum esculentum*	10 percent	8	890	12.0
Lacy Phacelia**	*Phacelia tanacetifolia*	10 percent	8	13,750	1.0
Annual Sunflower	*Helianthus annuus*	8 percent	6.4	2,810	3.0
TOTALS		**100 percent**	**80**		**26.5**

**May attract pests of brassica crops (e.g., broccoli, arugula, etc.) like flea beetles and harlequin bug. Omit from the mix if this is a concern.*

***Reported to support lygus bug pests. Consider omitting or locating away from lygus-susceptible crops like strawberries.*

Perennial Mix, Western United States

ADVANTAGES: The expected life Span of this mix is at least five years, and after establishment it won't require irrigation in most locations.

DISADVANTAGES: Perennial wildflowers planted from seed are unlikely to bloom for at least a full year. Perennial wildflowers are also more sensitive to proper planting time and seeding depth, so surface planting is often best.

Common Name	Scientific Name	Percent of Mix	Seeds per Square Foot	Seeds per Ounce	Ounces Needed
Sulfur Buckwheat	*Eriogonum umbellatum*	5 percent	4	13,125	0.25
Yarrow	*Achillea millefolium*	10 percent	8	171,875	0.25
Perennial Blanketflower	*Gaillardia aristata*	15 percent	12	11,625	1.25
Alfalfa	*Medicago sativa*	20 percent	16	14,190	1.5
Idaho Fescue	*Festuca idahoensis*	25 percent	20	28,125	1.0
Blue Wild Rye	*Elymus glaucus*	25 percent	20	7,500	3.50
TOTALS		**100 percent**	**80**		**7.75**

*Even in dry western regions, annual sunflower (*Helianthus annuus*) and plains coreopsis (*Coreopsis tinctoria*) are easy to grow as insectary plants.*

7 Hedgerows

- Hedgerows surround the farm with long-lived trees and shrubs.
- European hedgerows have supplied benefits to farms for thousands of years.
- Research demonstrates that hedgerows export pest control into nearby crop fields.
- While the initial investment can be significant, hedgerows pay for themselves quickly with direct economic benefits.

HEDGEROWS ARE LINEAR ROWS of flowering shrubs and small trees, sometimes with wildflowers and grasses in the understory. They're typically located along property boundaries, fence lines, roads, and as barriers to physically separate crop fields. Hedgerows have long been recognized for their many farm benefits, including protection from wind and drifting snow, providing habitat for wildlife and pollinators, capturing runoff from agricultural lands, and providing renewable sources of firewood, fruits, and herbs.

The hedgerow concept is thought to be more than four thousand years old, having originated in Europe, where the remnants of cleared woodland were often used to mark property boundaries and provide living fences for livestock. Hedgerows more than seven hundred years old still exist in parts of Britain! European hedgerows are characterized by various styles associated with their respective regions: they are sometimes constructed on top of earthen banks and supported with stacked stones, and ditches often run along their entire length.

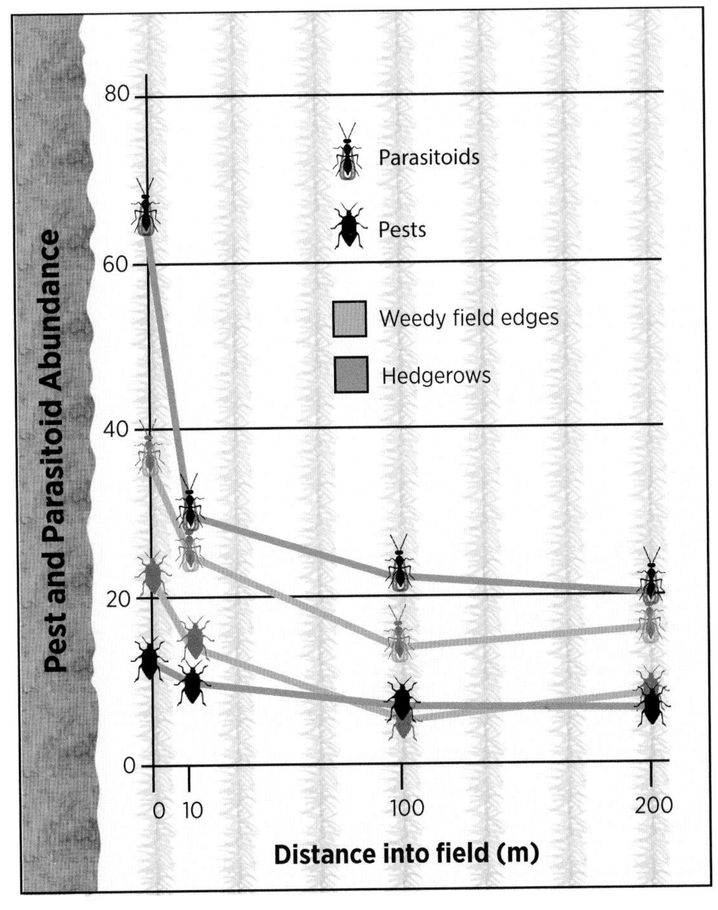

Parasitoid and pest abundance on sticky card traps in tomato fields adjacent to native hedgerows, versus tomato fields adjacent to weedy field edges. In tomato fields adjacent to hedgerows, parasitoids were more abundant and pests less abundant.

SURE, BUT HOW MUCH DOES IT COST?

If you're like many farmers, the upfront costs of establishing hedgerows and other beneficial insect habitat can seem daunting. As part of a multiyear research project at the University of California Berkeley, Drs. Lora Morandin and Claire Kremen and Rachael Long evaluated the initial costs of establishing a farm hedgerow in relation to the long-term benefits of pollination and pest management. They found that for tomato crops, these combined benefits offset the hedgerow installation costs within 10 years. Establishment costs were offset even sooner if the hedgerow was financed with cost-share assistance from the USDA-NRCS.

Because hedgerows require relatively low levels of maintenance after planting, and because, as hedgerows in Europe demonstrate, they may last for hundreds of years, they are a relative bargain for farmers with an interest in long-term conservation.

CREATING A HEDGEROW

Hedgerows can be created on level ground or on berms, which may provide greater screening and windbreak benefits. To build a berm, use a single-blade plow to create a long, linear mound, then turn around and plow in the opposite direction to increase the size of the mound.

When planting the hedgerow it is important to install irrigation, such as drip tape, to water the new transplants until they can fend for themselves.

Hedgerows, increasingly popular in the United States, are often constructed along roads, where they can help reduce gravel dust, or along irrigation ditches to help stabilize the soil. The use of flowering trees and shrubs in hedgerows can help maintain those primary conservation functions while at the same time support an abundance of beneficial insects.

CHOOSING PLANTS

Native flowering vines can also be incorporated into hedgerows, along with various native grasses and wildflowers; consider shade-tolerant species for north-facing sides of hedgerows. In some cases, where the availability of native plants is limited or cost prohibitive, or for droughty, exposed sites unfavorable to native species, noninvasive exotics can also be used. For example, rosemary (*Rosmarinus officinalis*) and Russian sage (*Perovskia atriplicifolia*) are two widely available nonnative plants that flower prolifically, tolerate dry, exposed locations, and form small shrubs in warm climates.

GETTING STARTED

Most hedgerow plants are established from transplants or live-stake cuttings. Such transplants usually need supplemental irrigation for the first two years after transplanting, along with regular weeding and sometimes protection from browsing deer or rodents. Transplanted shrubs may also require several years of establishment before they begin flowering.

If you want immediate pollen and nectar resources, you can "jump-start" a hedgerow by interseeding wildflowers between the transplanted shrubs or in rows adjacent to the sides of the hedgerow. As the shrubs mature, they will crowd out some of the area occupied by wildflowers, but begin to produce more of their own flowers.

See the case study *Hedgerows on California Central Valley Farms*, page 113. See also part 5 for some common native flowering trees and shrubs used as hedgerow plants in various regions.

A heavy layer of mulch (such as wood chips) will help reduce transplant stress by preventing weed competition and conserving soil moisture.

After several years, irrigation to the hedgerow should be gradually reduced and eventually eliminated altogether. To increase the wildlife value of hedgerows, you can coppice-cut shrubs to encourage dense resuckering, leave pruned material in the hedge as nesting material, and plant wildflowers in the understory.

Long-term hedgerow management usually consists of removing invasive weeds, or cutting back large trees that shade other hedgerow species. Larger hedgerow species in Europe are sometimes cut back to the ground and allowed to resprout (called coppicing) to produce multiple bushy stems. Another practice, called hedge-laying, involves cutting horizontally most of the way through upright trunks, then pushing the trunks, still partially attached, over at an angle in line with the hedgerow. These are held in place by stakes and bound along the top with long, supple stems, usually of hazel. New growth from the stumps and laid trunks results in thicker hedgerow structure and fills in gaps where other shrubs may have died.

Finally, while hedgerows can be used to provide habitat for beneficial insects, they can also be used to protect beneficial insects from pesticide spray. Hedgerows consisting of dense evergreen trees like spruce and arborvitae will do little to support most insect predators and parasitoids. However, those plants can provide wind screening and reduce pesticide drift from adjacent cropland. Such pesticide screening may be an especially important consideration for organic producers adjacent to conventional farms. For details on the creation of hedgerows for pesticide drift reduction, see chapter 11.

As with native plant field borders, hedgerows are a recognized Conservation Practice Standard for the USDA Natural Resources Conservation Service (NRCS), and you may qualify for financial assistance to install them. Your local NRCS office and the Xerces Society can provide technical assistance in hedgerow design.

Installing a New Hedgerow

Hedgerows are relatively easy to establish, and anyone who has ever planted a tree or shrub already knows the basics. By focusing on some of the finer points noted here, however, you will ensure that your hedgerow is optimized for beneficial insects.

Planning for Success

When you select trees and shrubs for planting, consider the future mature size of your hedgerow. If your hedgerow will be directly adjacent to a crop field, you may want to focus on a single row of small shrubs that won't significantly shade the nearby crops. If you have a wider area to work with, you might consider a multirow, multitiered hedgerow that includes a row of taller trees running parallel to another row of smaller shrubs and herbaceous plants. Similarly, wider hedgerows can be established in double staggered rows (a zigzag configuration) to create a bushy, fairly dense wall of vegetation.

Once you have your primary tree and shrub species in mind, you'll want to measure the size of the area to estimate how many plants you need to fill the space. At this stage you might also want to place survey flags or stakes along the planting area to help you get a sense of what the final hedgerow will look like.

In general, to create a thick, wall-like hedgerow, you should space the individual plants 2 to 10 feet (0.6 to 3 m) apart depending upon the mature size of the plant. Shrubs under 6 feet (1.8 m) tall should be planted close together, and larger trees farther apart. Any herbaceous plants that you include can be spaced even closer, at distances of 1.5 to 3 feet (0.5 to 1 m) apart.

During the preparation phase, you might also think about whether you want to plant the hedgerow on flat ground or if you first want to create a berm to serve as the hedgerow base. Hedgerows with berm bases can provide greater windbreak and screening benefits, and they are considered the norm in some parts of Europe, where some bermed hedgerows are

This newly planted hedgerow does not look impressive yet, but with careful nurturing it will become a conservation feature that could last for hundreds of years.

The best hedgerows for wildlife and beneficial insects include plants of diverse sizes and shapes, including taller trees, dense shrubs, understory, and low-growing edge plants.

hundreds of years old and have survived two world wars. Hedgerow berms are typically 3 feet (1 m) wide and high, and are created using soil excavated from the sides of the berm; this creates a parallel ditch on both sides of the hedgerow to assist with drainage. Field stones are sometimes added to hedgerow berms as well, adding height and additional shelter for beneficial insects. While creating a bermed hedgerow base requires more initial work, the major benefit is that the resulting hedgerows tend to be more difficult to dismantle, and thus may be more likely to survive for multiple generations, providing long-term habitat for beneficial wildlife.

Before planting, you will also want to prepare the area by removing any existing vegetation, especially aggressive perennial weeds. Where weed pressure is low, such as in areas that have been regularly mowed for several seasons, consider preparing the site for planting by just mowing again at a very low height to make planting easy. If weed pressure is fairly intense, such as in areas that have been unmanaged, with annual weeds that have gone to seed for several years or that are now

HEDGEROWS HAVE A PLACE AT HOME

In the context of this book, hedgerows designed for beneficial insects generally take the form of long rows of diverse shrubs, sometimes extending for a mile or more along the edges of farm fields and roads. Their role is to provide habitat that supports pest control, but also to provide corridors for wildlife, mark field and property boundaries, decrease trespassing, reduce the drift of dust or other pollutants, and beautify agricultural lands. However, the same goals can be achieved even around urban or suburban properties.

The Xerces Society's Mace Vaughan has experimented with hedgerows at his own home on a typical 50-by-100-foot (15 by 30 m) city lot in Portland, Oregon, creating a beautiful landscape feature that achieved all the goals listed above, even though the hedgerow was only 50 feet long.

To start, Mace created a berm and low stone wall along a sidewalk. Into the berm he planted snowberry, ocean spray, meadowsweet, currant, and mockorange, which have grown up together. After two years of watering to ensure good establishment, the result is a low wall of shrubs with blooms covered in beneficial insects from spring to summer, a private front yard, and an attractive visual screen.

Mace's simple concept can be applied to any situation in which you would like to have a soft break between two properties, increase privacy from public right-of-ways, or simply create a low-maintenance habitat for beneficial insects. The urban hedgerow may be only 25 or 50 feet (8 or 15 m) long, but it will provide habitat and a beautiful landscape feature where you can watch flower flies and parasitic wasps forage as they take a break from attacking the pest aphids on your roses.

dominated by tall perennial weeds or invasive shrubs, you might need to spray a planting strip with herbicide or use a brush hog to clear the existing vegetation. You can also solarize weedy areas by very closely mowing the hedgerow site and laying down UV-stabilized greenhouse plastic for several weeks in midsummer; bury the edges of the plastic to prevent airflow beneath the sheet.

As a final but critical step, you will need to plan for irrigating your hedgerow transplants immediately after planting. In most regions, transplanted shrubs will need 1 inch (2.5 cm) of water per week for the first two years after installation. After that you can taper off the irrigation, and by the fourth or fifth year after planting, your hedgerow shouldn't need to be watered at all except during severe drought. With this in mind, some farmers prefer to install drip irrigation or sprinklers in their hedgerows. If this isn't an option, make sure you have at least a long hose that will reach the entire length of the planting site. Drip irrigation to the base of each plant is ideal, to help establish the hedgerow plants without also watering surrounding weeds.

Planting Methods

Before planting begins, stage your transplants in the planting area. This will give you a last chance to look at the overall arrangement, and to make any last-minute changes in the way the plants are organized. If the weed pressure is high, you might also want to install landscape fabric at this point, In this case, you will plant into holes cut through the fabric, and typically then cover the entire understory with mulch. While this practice may be highly effective for weed control, it does reduce soil access for ground-dwelling beneficial insects and other wildlife. Hedgerows should be installed without landscape fabric when possible.

When you are ready to plant, a regular shovel is usually adequate for small nursery transplants, but power augers and mechanical tree spades can also be rented to make holes rapidly or to install larger plants. Be careful not to damage the root mass during planting, especially if it is loose and fragile. Also be sure to place the transplant in a hole without compacted sides, and at the proper depth, which is usually just above the first major root so that the crown is at the soil surface. If underground rodent damage is likely, you might want to install wire root cages during the planting process. Similarly, you may need trunk guards to protect plants from aboveground browsing or antler damage by deer.

Large-scale hedgerow construction can require careful design and specialized materials such as erosion control wattles, truckloads of mulch, and irrigation systems. While this may seem like a large investment, research demonstrates that farmers can quickly recoup those costs through reduced pest damage and enhanced insect pollination. USDA conservation program funds are also often available to help offset the initial investment.

In most regions, you can plant hedgerows anytime you can work the ground, but avoid planting during periods of hot, dry, or windy weather. Regardless of when you plant, make sure the transplants are thoroughly irrigated immediately after they are placed in the ground. To prevent transplant shock, fill the holes with water and allow them to drain prior to planting. To help retain moisture and reduce weed competition, you should also spread mulch immediately after planting, but avoid pushing it up against the trunk of the transplant. For mulch, you can use wood chips, bark dust, weed-free straw, nut shells, or other suitable materials available in your area.

Post-Planting Care

Weed control in the first and second years after planting is one of the most important management steps you can take to ensure the survival of your hedgerow. If you prepared the site well before planting, you'll expend less effort on weeding later. Depending on your weeds and your available tools, you might use mowing, hand hoeing, or spot spraying with herbicides as your primary weed control strategy. It is especially important to prevent weeds from going to seed during the first few years after planting to give your hedgerow a good chance for success.

If deer, rabbits, or ground squirrels are around, you might need to control them or fence them out. However, be sure to remove tree guards, fencing, or other barriers that could interfere with plant growth as soon as possible. In most cases, transplants can be removed from irrigation by the end of the second year after planting.

In the long term, you might need to prevent hedgerows from growing into adjacent fields or roadsides. You might also need to cut back large trees that shade out other hedgerow species. Depending on your goals, and on the individual species, some larger hedgerow plants can be coppiced, or periodically cut back to a stump and allowed to resprout to produce multiple bushy stems. Never, however, prune hedgerow plants during wildlife nesting seasons, which usually occur in spring and summer. After it is established, ideally no more than 30 percent of your hedgerow should be disturbed in any one year to ensure sufficient undisturbed and continuous habitat for beneficial insects.

Before installing hedgerow transplants, it is a good idea to stage them along the planting area, thinking about how they will blend together at their mature height, and ensuring that they line up with drip irrigation emitters.

With active management and selective transplanting, neglected fence-line areas like this can be shaped into attractive and functional hedgerow systems.

Revitalizing Old Fencerows

Informal hedgerows are already found along the property boundaries and fencerows of many farms. Sometimes these hedgerows are the result of neglect, sometimes they grow up because equipment can't operate right up next to the fence, and sometimes they exist because neither neighbor wants to impose on the other by actively managing a space. These unplanned and unmanaged hedgerows may provide huge benefits to wildlife, and are sometimes the only wildlife corridors that exist in intensively farmed areas.

If you have an old fencerow that has grown up into volunteer trees and shrubs, you have a potential beneficial insect goldmine that may already be providing pest control for the farm. More typically, however, unmanaged fencerows contain a combination of good plants and invasive weeds. With some diligent work, you can usually turn these unmanaged areas into something great.

Focus first on identifying what plant species are present in your shrubby fencerow. Anything that is an invasive weed or a potential host of pest insects should be killed by girdling, spot spraying, or removing with a weed wrench. If it won't be in the way, feel free to leave the old brush from the weedy shrubs piled in the fencerow, where it will provide shelter for beneficial insects. Then assess the condition of the fence itself, and decide whether you need to repair it, or whether you can possibly remove it.

Next, if you can get irrigation water out to the fencerow, consider immediately replanting more desirable species in the spots where you removed weedy plants. If you can populate bare ground with desirable shrubs, you'll create less space for the weedy ones to grow back. You can also sometimes fill in bald spots next to desirable shrubs by carefully bending flexible side shoots down, and training them to grow along the ground. Many shrubs will form adventitious roots if their stems are staked to the ground and buried with a small amount of soil; this causes them to grow outward and send up new shoots or suckers. If you are concerned about shrubs growing into and damaging an existing fence, you should plant them as far away from the wires as possible, and consider creating a small path next to the fence that you can keep clear with a chainsaw when you need to.

If larger trees are present in your fencerow, think about whether they overly shade adjacent crops or, conversely, provide beneficial shade for farmworkers or livestock. Similarly, identify any weedy species like tree-of-heaven (*Ailanthus altissima*) that shower seed everywhere and create problematic seedlings. If a tree is a problem, but it's not in a spot likely to be a hazard, consider girdling it and leaving it standing for beneficial wildlife, such as raptors, to use.

The last factor to consider is how your fencerow management will affect relations with your neighbors. This book may provide a starting point for that dialogue, but working with your neighbors to identify and meet their goals for the fencerow is essential. If you can point out how your improvements will reduce pests, provide hunting opportunities for game such as pheasant or quail, and make both your farms more attractive, it might encourage your neighbors to help you, or at least to give you the green light to improve the fencerow on your own.

Sample Hedgerow Plant Mixes

The following lists represent regional examples of what you might plant to create your own native shrub hedgerow. Most of these species are commonly available from local native plant nurseries.

Northeastern United States

Common Name	Scientific Name	Mature Height (ft)	Bloom Time	Notes
TALL HEDGEROW PLANTS				
Black Willow	*Salix nigra*	100	Early spring	Prefers wet sites; mature trunks tend to collapse but remain growing as fallen trees
Pussy Willow	*Salix discolor*	30	Early spring	Prefers wet sites; suitable for coppice cutting
Basswood	*Tilia americana*	100	Spring	Tall tree at maturity
Cockspur Hawthorn	*Crataegus crus-galli* *	30	Spring	Tolerant of dry or wet soils
Chokecherry	*Prunus virginiana* *	25	Spring	Tolerant of dry or wet soils
Downy Hawthorn	*Crataegus mollis* *	30	Spring	
False Indigo Bush	*Amorpha fruticosa*	15	Late spring	Considered weedy in some riparian areas
Common Buttonbush	*Cephalanthus occidentalis*	15	Summer	Prefers wet sites
Autumn Willow	*Salix serissima*	10	Autumn	Prefers wet sites
SHORT HEDGEROW PLANTS				
New Jersey Tea	*Ceanothus americanus*	6	Early summer	Heavily browsed by deer
Carolina Rose	*Rosa carolina*	5	Summer	
Steeplebush	*Spiraea tomentosa*	4	Summer	
Swamp Rose	*Rosa palustris*	8	Summer	Prefers wet sites
Woodland Sunflower	*Helianthus divaricatus*	5	Summer	Herbaceous understory plant found in thickets
Calico Aster	*Symphyotrichum lateriflorum*	3	Autumn	Herbaceous understory plant found in thickets

**May support plum curculio or apple maggot. Consider omitting or locating away from susceptible crops like apples or plums.*

Note: When planting a hedgerow that includes species that are very tall at maturity, consider locating them 100 feet apart or more to avoid shading the hedgerow, or locating them in field corners at the end of a hedge where they can function as a small grove or woodlot.

Southeast

Common Name	Scientific Name	Mature Height (ft)	Bloom Time	Notes
TALL HEDGEROW PLANTS				
Coastal Plain Willow	*Salix caroliniana*	25	Early spring	Prefers wet sites
Buckwheat Tree	*Cliftonia monophylla*	30	Early spring	
Chickasaw Plum	*Prunus angustifolia**	12	Early spring	
False Indigo	*Amorpha fruticosa*	15	Spring	Prefers wet sites
Common Buttonbush	*Cephalanthus occidentalis*	15	Summer	Prefers wet sites
Eastern Baccharis	*Baccharis halimifolia*	10	Autumn	Poisonous to livestock; separate male and female plants
SHORT HEDGEROW PLANTS				
Carolina Rose	*Rosa carolina*	5	Summer	
Swamp Rose	*Rosa palustris*	6	Summer	
Tall Ironweed	*Vernonia angustifolia*	6	Summer	Tall, bushy herbaceous perennial; prefers well-drained sandy soil
Starry Aster	*Silphium asteriscus*	6	Summer	Tall, bushy herbaceous perennial; prefers well-drained sandy soil
Climbing Aster	*Ampelaster carolinianus*	8	Autumn	Vinelike plant, requires supporting vegetation

**May support plum curculio or apple maggot. Consider omitting or locating away from susceptible crops like apples or plums.*

Midwest

Common Name	Scientific Name	Mature Height (ft)	Bloom Time	Notes
TALL HEDGEROW PLANTS				
Black Willow	*Salix nigra*	100	Early spring	Prefers wet sites; mature trunks tend to collapse but remain growing as fallen trees
Pussy Willow	*Salix discolor*	30	Early spring	Prefers wet sites; suitable for coppice cutting
Basswood	*Tilia americana*	100	Spring	Tall tree at maturity
Cockspur Hawthorn	*Crataegus crus-galli**	30	Spring	Tolerant of dry or wet soils
Chokecherry	*Prunus virginiana**	25	Spring	Tolerant of dry or wet soils
Common Elderberry	*Sambucus canadensis*	10	Summer	
Common Buttonbush	*Cephalanthus occidentalis*	15	Summer	Prefers wet sites
SHORT HEDGEROW PLANTS				
Leadplant	*Amorpha canescens*	3	Early summer	
New Jersey Tea	*Ceanothus americanus*	6	Early summer	Heavily browsed by deer
Carolina Rose	*Rosa carolina*	5	Summer	
Steeplebush	*Spiraea tomentosa*	4	Summer	
Woodland Sunflower	*Helianthus divaricatus*	5	Summer	Herbaceous understory plant found in thickets
Calico Aster	*Symphyotrichum lateriflorum*	3	Autumn	Herbaceous understory plant found in thickets

**May support plum curculio or apple maggot. Consider omitting or locating away from susceptible crops like apples or plums.*

Northern Plains

Common Name	Scientific Name	Mature Height (ft)	Bloom Time	Notes
TALL HEDGEROW PLANTS				
Pussy Willow	*Salix discolor*	30	Early spring	Prefers wet sites; suitable for coppice cutting
Chokecherry	*Prunus virginiana**	25	Spring	Tolerant of dry or wet soils
Fleshy Hawthorn	*Crataegus succulenta**	20	Spring	
False Indigo Bush	*Amorpha fruticosa*	15	Late spring	Tolerant of dry or wet soils
Scarlet Elderberry	*Sambucus racemosa*	15	Summer	
SHORT HEDGEROW PLANTS				
Leadplant	*Amorpha canescens*	3	Early summer	
Buffaloberry	*Shepherdia argentea*	8	Early summer	
Prairie Rose	*Rosa arkansana*	3	Summer	
White Spirea	*Spiraea alba*	2	Summer	
Maximilian Sunflower	*Helianthus maximiliani*	8	Late summer	Tall herbaceous perennial; incorporates well in hedgerow edges
Old Field Goldenrod	*Solidago nemoralis*	3	Autumn	Herbaceous perennial; incorporates well in hedgerow edges

**May support plum curculio or apple maggot. Consider omitting or locating away from susceptible crops like apples or plums.*

Southern Plains

Common Name	Scientific Name	Mature Height (ft)	Bloom Time	Notes
TALL HEDGEROW PLANTS				
Black Willow	*Salix nigra*	100	Early spring	Prefers wet sites; mature trunks tend to collapse but remain growing as fallen trees
Chickasaw Plum	*Prunus angustifolia**	12	Early spring	
Chokecherry	*Prunus virginiana**	25	Spring	Tolerant of dry or wet soils
False Indigo Bush	*Amorpha fruticosa*	15	Late spring	
Common Buttonbush	*Cephalanthus occidentalis*	15	Summer	Prefers wet sites
Black Elderberry	*Sambucus nigra*	10	Summer	
SHORT HEDGEROW PLANTS				
Inland Ceanothus	*Ceanothus herbaceus*	2	Early summer	
Leadplant	*Amorpha canescens*	3	Early summer	
Prairie Rose	*Rosa arkansana*	3	Summer	
Showy Milkweed	*Asclepias speciosa*	5	Summer	Tall herbaceous perennial; incorporates well in hedgerow edges
Maximilian Sunflower	*Helianthus maximiliani*	8	Late summer	Tall herbaceous perennial; incorporates well in hedgerow edges
Giant Goldenrod	*Solidago gigantea*	8	Autumn	Tall herbaceous perennial; incorporates well in hedgerow edges

**May support plum curculio. Consider omitting or locating away from susceptible crops like apples or plums.*

Intermountain West

Common Name	Scientific Name	Mature Height (ft)	Bloom Time	Notes
TALL HEDGEROW PLANTS				
Scouler's Willow	*Salix scouleriana*	20	Early spring	Prefers wet sites; can be maintained as a tree or shrub based on pruning
Chokecherry	*Prunus virginiana**	25	Spring	Tolerant of dry or wet soils
Black Hawthorn	*Crataegus douglasii**	15	Spring	
Ocean Spray	*Holodiscus discolor*	12	Summer	
Black Elderberry	*Sambucus nigra*	10	Summer	
SHORT HEDGEROW PLANTS				
Buffaloberry	*Shepherdia argentea*	8	Early summer	
Mountain Ceanothus	*Ceanothus velutinus*	8	Summer	Heavily browsed by deer and elk
Nootka Rose	*Rosa nutkana*	9	Summer	Prefers damp soils
Desert Sweet	*Chamaebatiaria millefolium*	6	Summer	Very drought tolerant
Sulfur Buckwheat	*Eriogonum umbellatum*	2	Late summer	Very drought tolerant
Canada Goldenrod	*Solidago canadensis*	3	Autumn	Tall herbaceous perennial; incorporates well in hedgerow edges

**May support apple maggot. Consider omitting or locating away from apples.*

Southwest

Common Name	Scientific Name	Mature Height (ft)	Bloom Time	Notes
TALL HEDGEROW PLANTS				
Arroyo Willow	*Salix lasiolepis*	35	Early spring	Prefers wet sites; suitable for coppice cutting
Chokecherry	*Prunus virginiana*	25	Spring	Tolerant of dry or wet soils
Blue Elderberry	*Sambucus nigra cerulea*	15	Spring	
Emory Baccharis	*Baccharis emoryi*	10	Autumn	
Mulefat	*Baccharis salicifolia*	10	Autumn	
SHORT HEDGEROW PLANTS				
Desert Ceanothus	*Ceanothus greggii*	8	Spring	
Coffeeberry	*Rhamnus californica*	5	Spring	Very drought tolerant
Desert Rose	*Rosa stellata*	3	Late spring	
Buffaloberry	*Shepherdia argentea*	8	Early summer	
Desert Sweet	*Chamaebatiaria millefolium*	6	Summer	Very drought tolerant
Fendler's Ceanothus	*Ceanothus fendleri*	8	Summer	
Sulfur Buckwheat	*Eriogonum umbellatum*	2	Late summer	Very drought tolerant
Canada Goldenrod	*Solidago canadensis*	3	Autumn	Tall herbaceous perennial; incorporates well in hedgerow edges

California

Common Name	Scientific Name	Mature Height (ft)	Bloom Time	Notes
TALL HEDGEROW PLANTS				
Toyon	*Heteromeles arbutifolia*	10	Spring	Alternate host for fire blight
Blue Elderberry	*Sambucus nigra cerulea*	15	Spring	
Hollyleaf Cherry	*Prunus ilicifolia*	15	Late spring	
Common Buttonbush	*Cephalanthus occidentalis*	15	Summer	Prefers wet sites
Coyotebush	*Baccharis pilularis*	10	Autumn	Very drought tolerant
SHORT HEDGEROW PLANTS				
California Lilac	*Ceanothus* spp.	6	Spring	
Coffeeberry	*Rhamnus californica*	5	Spring	Very drought tolerant
California Rose	*Rosa californica*	8	Spring	
Narrowleaf Milkweed	*Asclepias fascicularis*	3	Summer	Herbaceous perennial; incorporates well in hedgerow edges
Pacific Aster	*Symphyotrichum chilense*	3	Autumn	Tall herbaceous perennial; incorporates well in hedgerow edges
Canada Goldenrod	*Solidago canadensis*	3	Autumn	Tall herbaceous perennial; incorporates well in hedgerow edges

Pacific Northwest

Common Name	Scientific Name	Mature Height (ft)	Bloom Time	Notes
TALL HEDGEROW PLANTS				
Sitka Willow	*Salix sitchensis*	20	Early spring	Prefers wet sites; can be maintained as a tree or shrub based on pruning
Scouler's Willow	*Salix scouleriana*	20	Early spring	Prefers wet sites; can be maintained as a tree or shrub based on pruning
Chokecherry*	*Prunus virginiana*	25	Spring	Tolerant of dry or wet soils
Red Stem Ceanothus	*Ceanothus sanguineus*	10	Spring	Slow growing
Cascara	*Frangula purshiana*	35	Spring	Tall at maturity
Ocean Spray	*Holodiscus discolor*	12	Summer	
Coyotebush	*Baccharis pilularis*	10	Autumn	Very drought tolerant
SHORT HEDGEROW PLANTS				
Buckbrush	*Ceanothus cuneatus*	8	Summer	Slow growing
Douglas Spirea	*Spiraea douglasii*	6	Summer	
Nootka Rose	*Rosa nutkana*	9	Summer	Prefers damp soils
Showy Milkweed	*Asclepias speciosa*	5	Summer	Herbaceous perennial; incorporates well in hedgerow edges
Hall's Aster	*Symphyotrichum hallii*	4	Autumn	Herbaceous perennial; incorporates well in hedgerow edges

**May support apple maggot. Consider omitting or locating away from apples.*

CASE STUDY

Hedgerows on California Central Valley Farms

IN THE HEART of the Sacramento Valley, California, Yolo County growers are at the forefront of an effort to plant hedgerows on their farms. For example, during the past four years, 17 miles of hedgerows were planted in Yolo County alone, representing 25 percent of the total hedgerow miles planted in California and 8 percent nationally during this time. These corridors of vegetation primarily surround field crops, including canning tomatoes, wheat, alfalfa, and other field crops, worth over $200 million annually.

There are many reasons that farmers plant hedgerows on their farms. Some like the biodiversity hedgerows bring to intensively farmed areas. Birds and other wildlife frequent the hedgerows, providing wildlife viewing for some and game hunting for others. For example, in a recent study in Yolo County, the presence of hedgerows on farms tripled the abundance and doubled the diversity of birds, without attracting additional bird pests to adjacent crops. The flowering shrubs and berries produced in the hedgerows are also visually attractive throughout the year, especially after crops are harvested and the ground is disked and bedded in preparation for the next year's crop. Farmers are also interested in hedgerows to replace weedy field margins as well as for windbreaks and to prevent soil erosion. All of this in turn helps improve air and water quality.

California's Central Valley is home to most of the beneficial insect hedgerows in the U.S., but the concept is appropriate for most of North America — and a vastly underutilized conservation practice.

There is also significant interest in the enhanced pollination and pest control that beneficial insects thriving in hedgerows can provide to adjacent crops. Long-term studies in Yolo County have shown that hedgerows and native grasses attract a wide variety of insect predators and parasitoids that feed on the nectar and pollen of flowering plants. These natural enemies include ladybird beetles, big-eyed and minute pirate bugs, and parasitic and predatory wasps and flies. These beneficial insects have also been shown to move long distances into adjacent crops where they help suppress pests. Although some pests are found in hedgerows, the flowering shrubs support a proportionately far greater population of beneficial insects than is found in weedy field margins.

Most growers elect to plant native California shrubs and perennial grasses in hedgerows, including California lilac (*Ceanothus* spp.), toyon (*Heteromeles arbutifolia*), elderberry (*Sambucus* spp.), coffeeberry (*Rhamnus californica*), coyote brush (*Baccharis pilularis*), redbud (*Cercis occidentalis*), purple needlegrass (*Nassella pulchra*), oniongrass (*Melica californica*), and blue wild rye (*Elymus glaucus*). These plants are drought tolerant and thrive well, once established, in California's hot, dry summers with minimal irrigation in most soils. Most of the hedgerows are planted along field margins in areas that cannot be farmed, so no land is taken out of production. These include plantings along streams, canals, fence lines, and areas where differences in field height occur as a result of land leveling. Most hedgerows are linear, with a row of shrubs bordered by perennial grasses, as these are easiest to care for with large-scale farm equipment, including mowers for weed control.

As farmers in California learn about hedgerows and become aware of their many benefits, interest in them and in corridors of vegetation surrounding farms is continually growing. Numerous field days, workshops, field trips, NRCS cost-share programs, and the experience of local conservationists are making this possible. Thriving hedgerows in intensively farmed areas of California's Central Valley add rich biodiversity to its landscape and potential economic benefits to its farms.

— RACHAEL LONG, University of California Cooperative Extension

Cover Crops

8

COVER CROPS ARE TEMPORARY or permanent plantings of ground cover on fallow crop fields, between rows of berry crops, or in the understory of vineyards and orchards. Like hedgerows, cover cropping can have multiple conservation objectives, including reducing erosion, improving soil fertility, preventing weed growth, breaking pest and disease cycles, and providing pollen, nectar, and shelter for beneficial insects.

- Cover crops provide multiple benefits to farms.
- Both temporary and perennial cover cropping systems are available to support any type of farming.
- Multispecies cover crops can have advantages over single species for beneficial insects.
- Cover cropping is easy to do.

Species Selection

Common cover crops include various legumes, grasses, and brassicas. Legumes are widely recognized for their contribution to soil fertility as green manure crops. Grasses, in contrast, are noted for their ability to capture excess soil nutrients, prevent weed growth, and reduce erosion. Brassicas are commonly used to absorb excess nutrients, alleviate soil compaction, and suppress soil pests such as nematodes. Additional cover crops, such as lacy phacelia and buckwheat, are used for various applications, including the support of beneficial insects. Depending on your objectives, you can seed several different cover crop species together to provide complementary benefits.

Legumes

Legume cover crops that provide pollen and nectar for beneficial insects include perennial species such as red, white, and alsike clovers (*Trifolium pratense, T. repens,* and *T. hybridum*), biennial species such as sweet clover (*Melilotus* spp.), and annual species like crimson clover (*T. incarnatum*), berseem clover (*T. alexandrinum*), cowpea (*Vigna unguiculata*), and hairy vetch (*Vicia villosa*). Other less commonly cultivated legume cover crops are available that may also be appropriate for certain climates. In some cases, native legumes are also available for cover crop purposes. Partridge pea (*Chamaecrista fasciculata*), an annual native in much of the eastern United States, attracts numerous beneficial insects and forms a thick canopy, making it a promising candidate for cover crop applications.

*Partridge pea (*Chamaecrista fasciculata*), an annual native legume of the eastern United States, is a promising candidate for cover crop applications.*

Many farmers are increasingly adopting multispecies cover crop mixes ("cover crop cocktails"). This cool-season cover crop in North Dakota includes vetch, radish, oats, turnip, Phacelia, *and several other species. These diverse mixes provide greater benefits to soil health and tend to support more insect diversity than can single-species cover crops.*

*Crimson clover (*Trifolium incarnatum*) is an annual legume, frequently used as a winter cover crop in warmer climates like the West Coast and southern United States.*

*Red clover (*Trifolium pratense*) is a perennial legume that is used for both annual and long-term cover-cropping applications, especially in cool climates.*

*Junegrass (*Koeleria macrantha*) is a tough native perennial grass that will form a sodlike mass when seeded at a high rate.*

*Little bluestem (*Schizachyrium scoparium*) is a drought-tolerant native bunch grass of eastern and midwestern North America that may integrate easily into unmowed vineyard and orchard understories, especially in warm, sunny locations.*

Grasses

Common grass cover crops include cereal rye, wheat, triticale, and oats, all of which can provide valuable overwintering cover and mulch layers for beneficial insects. These cover crops are often planted in fall and allowed to either winterkill (in the case of oats), or resume growth in the spring, when they are usually plowed under and replanted with cash crops. These same cool-season annual grasses are also often mixed with annual legumes like crimson clover (best in warm climates) or hairy vetch (for colder winter climates) to form a complementary cover crop mix that builds soil organic matter, adds nitrogen, and more effectively protects fallow ground from runoff. If allowed to go to flower, this same type of combined grass-legume cover can greatly increase benefits to predator and parasitoid insects by providing a complex of both flower resources and shelter.

This Washington State vineyard maintains an understory of native bunch grasses. These particular grape growers don't apply pesticides and rely entirely on wild beneficial insects for pest control.

In perennial crop systems, such as vineyards and orchards where the ground will not routinely be cultivated, some farmers maintain permanent ground covers of native grasses like Junegrass (*Koeleria macrantha*), side oats grama (*Bouteloua curtipendula*), little bluestem (*Schizachyrium scoparium*), blue wild rye (*Elymus glaucus*), bluebunch wheatgrass (*Pseudoroegneria spicata*), and Idaho fescue (*Festuca idahoensis*). Once established, these tough, drought-resistant grasses are highly effective at controlling weeds between rows and withstanding equipment traffic; they also provide permanent cover for beneficial insect species. When necessary, you can mow these native grasses periodically, or interplant them with perennial cover crop legumes (e.g., red clover) or native wildflowers to further support beneficial insects.

Annual grasses like oats, triticale, and rye combine perfectly with vetch, Phacelia, *and annual clovers to form a dense, complementary cover.*

Brassicas

Common brassica cover crops include several species of rapeseed (*Brassica napus*), mustards (*Brassica* spp. and *Sinapis* spp.), turnips (*Brassica rapa*), and forage radish, sometimes called oilseed or daikon radish (*Raphanus sativus*). A growing body of research demonstrates that brassicas not only support beneficial insects, but their roots also produce chemical compounds called glucosinolates that suppress weed growth, soilborne plant diseases, and nematodes. (If these are your goals, select brassica varieties with high concentrations of erucic acid, rather than low-acid varieties such as canola, for maximum pest suppression.)

*A relatively new cover crop, oilseed radish (*Raphanus sativus*) has a deep taproot that reduces compaction, increases water infiltration and, when allowed to bolt, provides excellent pollen and nectar for beneficial insects.*

Many brassicas are annual or biennial species, and will only flower, providing pollen and nectar resources for beneficial insects, if they are actively growing in the spring or early summer. This timing can be complicated by the fact that brassicas are often planted in the fall as a cool-weather cover crop and allowed to winterkill (many brassicas are not winter hardy and will die at around 25°F). Thus, if you want to plant brassicas to support beneficial insects, it is best to use them in spring-planted cover crop systems, or for year-round use in warmer climates. Brassicas can be interseeded with grasses and legumes (ideally at a low rate because of their competitiveness) to form a complementary cover crop. Finally, a few brassicas, such as black mustard (*Brassica nigra*), can have weedy tendencies, or they can be an alternate host for cabbage root maggot (*Delia radicum*), flea beetles, and several other pests, as well as clubroot disease of cruciferous vegetables. Consult a local vegetable crop expert if these are potential concerns.

Buckwheat

Buckwheat (*Fagopyrum esculentum*) is a broadleaf annual traditionally produced for seed as an alternative grain crop; it matures in only 30 to 45 days. This rapid growth allows it to be used as a smother crop on fallow fields to quickly

*Buckwheat (*Fagopyrum esculentum*) is a fast-growing nectar source in warm weather. Often the nectar, and consequently the insect activity, is most abundant early in the morning.*

outcompete annual weeds, so long as enough water is available. When allowed to mature, buckwheat flowers prolifically and provides an abundance of nectar to flower flies, parasitoid wasps, and other beneficial insects. Because of its rapid development, two sowings of buckwheat can be performed in a typical summer season. It does not grow vigorously in cool weather, however, so adjust your planting times according to local conditions. Buckwheat also produces large volumes of seed and can become temporarily weedy when allowed to self-sow. To control volunteer seedlings, if this is a concern, mow buckwheat before seeds fully develop.

Lacy Phacelia

Lacy phacelia (*Phacelia tanacetifolia*) is a native California wildflower that has been bred for uniformity in Europe and used extensively as an annual cover crop to suppress weeds and capture excess nitrogen in soil. Its use as a cover crop in the United States has steadily been increasing since the 1990s, and low-cost seed is now widely available from seed suppliers. Each lacy phacelia plant produces a stunning number of flowers that continue to open in succession over a period of several weeks. Staggered planting dates of phacelia can further extend the flowering period. Large numbers of bees, especially honey bees and bumble bees, are attracted to phacelia; flower flies and various beneficial wasps are also common flower visitors. Phacelia winterkills at 20°F, requiring spring planting in cool climates, but in warmer climates it can be sown in the fall for spring bloom.

Like buckwheat, phacelia can become weedy if allowed to self-sow, especially in warm climates. Mowing before seed maturation can reduce volunteer seedlings. Also note that some strawberry growers have reported lygus bug problems when lacy phacelia is planted nearby. We have not seen reports of lygus bug problems when phacelia is planted adjacent to other crops, so it appears this problem is not widespread. If you grow strawberries, however, you may wish to avoid using phacelia in your cover cropping system.

Many other cover crops are available for specific uses, and may be useful for providing pollen and nectar resources or shelter for beneficial insects. Consult your local seed suppliers and Cooperative Extension Service for more options.

*Lacy phacelia (*Phacelia tanacetifolia*), a native California wildflower, was first adopted as a cover crop in Europe and is now widely used across the United States. The volume of nectar this plant produces is amazing — attracting huge numbers of bees in addition to predator and parasitoid insects.*

Remember that if conservation biocontrol is your primary goal, it is essential to wait until most flowers have bloomed before tilling or mowing a cover crop. Cover cropping is a recognized conservation practice for the USDA NRCS and may qualify for financial incentives available through NRCS programs.

Establishing a Cover Crop

The following instructions represent several common winter and summer cover cropping approaches for various regions of the country. Keep in mind, however, that these samples are hardly exhaustive either in the species recommended, or the circumstances in which they could be used. For example, some farmers in the Upper Midwest have developed systems for interseeding vetch between rows of corn crops, and some vineyards in California are using buckwheat as a cover crop between rows of vines.

For more information about these and many other approaches to using cover crops, check out the book *Managing Cover Crops Profitably*, published by the USDA Sustainable

Regardless of region, cover crop systems can be designed to fit into most crop rotations and farm systems. Increasingly, USDA conservation programs are offering financial incentives for the adoption of cover cropping, and farmers are recognizing that cover cropping offers benefits that synthetic inputs can't provide.

Agriculture Research and Education (SARE) program. The book is available as a free download on the SARE website at www.sare.org.

In addition to the management information here, when establishing a cover crop you should be aware of residual herbicide issues and how they can affect cover crop establishment. Similarly, you should have a plan for terminating (killing) the cover crop when it reaches the desired stage of maturity. Common methods of terminating a cover crop include tilling, mowing, spraying with herbicide, and the use of specialized crimper-rollers that flatten the standing cover crop and crimp the stem to prevent regrowth, leaving a thick mulch cover on the ground. Finally, you should also think ahead about how you will manage the residue cover crops create, as well as any volunteer seedlings that emerge after the cover crop is removed. Most farmers who are experienced with using cover crops have simple solutions to these challenges (such as planting into cover crop residue with no-till seed drills), so it may be a good idea to seek out the advice of other farmers in your area who are already using cover crops as part of their farm system.

Before planting a cover crop, plan on how and when you will terminate its growth. Specialized crimper-rollers are one way to terminate cover crops, but no-till seed drills are needed to conduct follow-up planting through the thick residue.

This North Dakota sunflower field has understory cover crop rows consisting of buckwheat, cowpeas, turnips, and other species. Managed without irrigation, synthetic fertilizers, or insecticides — including insecticide-treated sunflower seed — the yields of this field compare to neighboring, conventionally managed sunflowers.

ESTABLISHING A COVER CROP

To establish a cover crop on a small scale, first clear away old crop debris, ideally by incorporating it into the soil, creating a smooth seedbed.

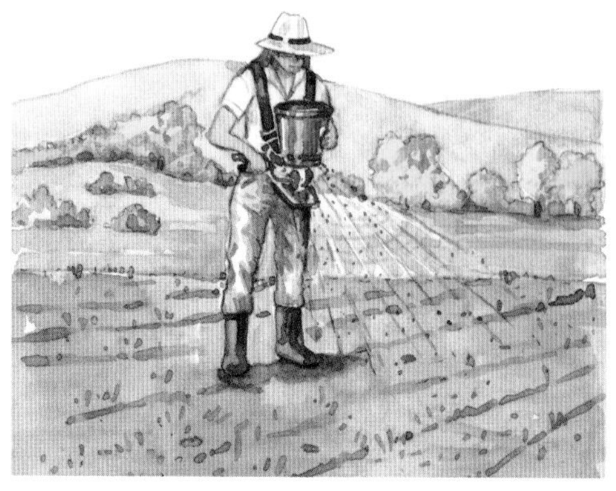

Except for large-seeded cover crops like buckwheat and grasses, most cover crop species can be planted by seeding directly on the soil surface.

Try to time your seeding just before seasonal precipitation to encourage fast germination.

Allowing the cover crop to flower before you terminate it is critical to encourage beneficial insects.

When clearing away the cover crop before planting a cash crop, leave strips of the cover crop standing on the field edge. As they continue to flower, they will continue to attract beneficial insects to the farm.

Winter Cover Crop

COLD CLIMATES (USDA HARDINESS ZONES 4–7)

Type	Seeding Rate	Sowing Season
Winter Rye	50 lbs/acre	Early fall (mid-September through mid-October)
Hairy Vetch	25 lbs/acre	
	TOTAL: 75 LBS/ACRE	

MODERATE CLIMATES (USDA HARDINESS ZONE 7 AND WARMER)

Type	Seeding Rate	Sowing Season
Winter Rye	65 lbs/acre	Early fall, 6 to 8 weeks before the first average frost, where frost is a threat.
Crimson Clover	20 lbs/acre	
	TOTAL: 85 LBS/ACRE	

Summer Cover Crop

ALL HARDINESS ZONES

Type	Seeding Rate	Sowing Season
Buckwheat	70 lbs/acre (broadcast); 50 lbs/acre (drill-seeded)	Spring and summer

SMALL-SCALE COVER CROPPING

Many gardeners overlook the practice of cover cropping even though it can offer them the same benefits that it provides to farmers who use it on a large scale. Along with supplying beneficial insects with pollen and nectar, flowering cover crops can build soil organic matter and capture excess nutrients. For example, in areas with a lot of winter precipitation, such as the Pacific Northwest, a winter cover crop of crimson clover in the vegetable garden can help reduce the compaction caused by daily rainfall, and can reduce the leaching of nutrients from the upper root zone of the soil.

Beyond traditional cover cropping, the home gardener may even consider managing an entire lawn as a permanent cover crop, with "no mow" turf grasses such as buffalo grass (*Bouteloua dactyloides*) and low-growing legumes such as white Dutch clover (*Trifolium repens*). Along with supporting beneficial insects, such cover crops can provide multiple benefits — for example, a lawn that doesn't require irrigation or fertilizer.

Traditional farm cover crops, like crimson clover (pictured here), offer the same benefits on a smaller scale to home gardens.

HOW TO

PLANT A WINTER COVER CROP

Rye and either hairy vetch (in cold climates) or crimson clover (in moderate climates) planted together will effectively control many cool-season weeds on fallow fields and build soil organic matter.

PEST CONTROL BENEFITS

- Rye may help suppress some pest nematodes and provides winter cover for predaceous beetles.
- When flowering in the spring, vetch and clover provide pollen and nectar sources for many beneficial insects. Clover especially attracts minute pirate bugs. Vetch and crimson clover may also host aphid populations that act as an early food source to increase beneficial insect populations.
- Hairy vetch may increase soybean cyst nematodes and should be avoided if this is a concern.

PLANTING METHOD

Optimal planting will place the rye seed deeper than the vetch or clover; thus, it is best to plant each species separately.

1. In early fall, establish the rye with a grain drill, placing the seed 1 to 1.5 inches (2.5 to 4 cm) deep.
2. Immediately after seeding the rye, overbroadcast the vetch or clover onto the field and roll it in with a cultipacker or scratch it in with a light harrow.
3. If a grain drill is not available, broadcast rye and vetch or clover together and roll or scratch the seed into the soil.

MANAGEMENT

- In late spring or early summer, terminate rye and vetch or clover cover crops by mowing, tilling, crimper-rolling, or herbicide.
- The longer the cover crop is allowed to persist, the greater benefits it will provide to predatory and parasitoid insects.
- Where possible, vegetable and field crop farmers might consider terminating the cover crop in strips and leaving live cover standing between crop rows to maintain beneficial insect habitat and provide erosion protection.

HOW TO

PLANT A SUMMER COVER CROP

The rapid growth and thick canopy of buckwheat make it highly effective at suppressing warm-season weeds. Also, buckwheat cover crops can build soil organic matter and capture excess soil nutrients.

PEST CONTROL BENEFITS

- The shallow white flowers excrete large amounts of nectar (especially in the morning hours) and support large numbers of beneficial insects, such as parasitoid wasps, flower flies, minute pirate bugs, and others. Depending on the location and planting time, buckwheat may flower for many weeks.
- Few pests and diseases are associated with buckwheat and it is not a known host for most major crop pests, although lygus bugs have occasionally been reported as infrequent pests of the crop.

PLANTING METHOD

1. Cultivate the soil in the spring as soon as it is dry enough to do so.
2. Wait at least two weeks to allow for existing organic matter to decompose, and then harrow to create a smooth seedbed.
3. Plant the buckwheat as soon as soil temperatures are at least 65°F (18°C); however, avoid planting immediately before any period of cool, wet weather.
4. If drill-seeding, place the seed no deeper than 1 inch (2.5 cm).
5. If broadcast seeding, scratch the seed into the soil surface with a light harrow.
6. One week after planting, broadcast additional seed over any gaps in the planting to suppress weed growth.

MANAGEMENT

- Buckwheat is a rapid-growing cover crop that will produce flowers around four to six weeks after planting.
- It can be established after early vegetable crops and terminated before fall crops are planted. It can also be allowed to flower and reseed itself, producing two bloom cycles in a typical summer.
- Buckwheat can be late-planted in the summer (after a grain crop) if enough time is available for it to develop before fall frosts.
- To maximize flowering, allow buckwheat to mature and be harvested for grain. If allowed to mature and not harvested, volunteer plants may be common. Those volunteers are easily controlled by mowing, tilling, crimper-rolling, or spraying.
- Buckwheat is not frost-tolerant and prefers warm weather and moderate irrigation of approximately ½ to 1 inch (1.2 to 2.5 cm) of water per week as it matures.

CASE STUDY

A Better Farm for Beneficials

AT WIND DANCER BLUEBERRY FARM in Grand Haven, Michigan, there has been a slow but steady transition toward a more biologically based approach to farming. Richard (RJ) Rant is incorporating plant diversity into his farm as part of a long-term movement toward more sustainable farming. As part of this movement, he is designing components to support beneficial insects while also improving soil nutrition and the health of his blueberry bushes.

To increase plant diversity within the rows of his 60-acre blueberry farm, RJ has planted different cover crops for different situations. He has planted Dutch and New Zealand white clovers (*Trifolium repens*) in areas that need to withstand farm traffic. Strawberry clover (*Trifolium fragiferum*), which can withstand wet conditions, was RJ's choice for low areas of the blueberry fields, to reduce standing water. He planted buckwheat (*Fagopyrum esculentum*) and crimson clover (*Trifolium incarnatum*) in alternate rows between the blueberry bushes. Because the crimson clover attracts many lacewings, aphids are no longer an issue on the farm. RJ is also testing several new insectary plantings between the rows, including one that will be planted with Michigan native wildflowers and grasses, and another to be planted with a mix of carrot (*Daucus carota*), chervil (*Anthriscus cerefolium*), coriander (*Coriander sativum*), clovers (*Trifolium* spp.), nasturtium (*Tropaeolum majus*), parsley (*Petroselinum crispum*), alyssum (*Lobularia maritima*), and yarrow (*Achillea millefolium*).

Under the blueberry bushes themselves, RJ is experimenting with a mix of crimson (*T. incarnatum*), white (*T. repens*), and alsike (*T. hybridum*) clovers that have short root systems that may outcompete weeds without competing with the shallow-rooted blueberry plants. The clovers fix nitrogen and provide low-pH root exudates for the acid-loving blueberry plants, and the flowers support beneficial insects, including bumble bees, that pollinate blueberry. The clovers also help build organic matter in RJ's sandy soils by providing a longer-release source of nutrients for the blueberry plants than fertilizers. Because the bushes are able to retain more moisture, RJ has been able to decrease his irrigation use. Although cover crops may be seen as an expensive initial investment, RJ has found that cover crops are more affordable than the combined costs of laying down wood chips and the extra fuel spent running his irrigation pump.

Looking ahead for other possibilities, RJ is now testing the use of several perennial cover crops, including yarrow, brassicas, turnips, oats, and common vetch, to combat weeds and support beneficial insects. Overall, RJ aims to put at least 5 percent of his property into habitat that will provide weed suppression, soil nutrients, and overlapping blooms through the season to help beneficial insects prosper on his farm. RJ views biodiversity as central to the improvement of his farm and the environment. After five years of this transition, his system is keeping pests in check, and has reduced the need for insecticides while still maintaining blueberry yields.

— DR. RUFUS ISAACS and BRETT BLAAUW, Michigan State University Department of Entomology

Crimson clover and other cover crops planted between rows of blueberry bushes in Michigan can attract beneficial insects while improving soil quality.

9 Conservation Buffers

- Buffers protect soil, water, and natural habitat from the impact of agriculture.
- With simple design modifications, buffers can be enhanced for beneficial insects.
- Financial and technical assistance for buffer construction is available through USDA conservation programs.
- Buffers can be scaled up or down to fit any size farm and any need.

BENEFICIAL INSECT HABITAT can be incorporated into farm systems in places other than fields. Those places include ditches and embankments that drain water from fields and roads, or areas of permanent vegetation that capture surface runoff and filter out sediment, excess nutrients, or agrochemicals. These landscape features are commonly called "buffers" or "filter strips."

Many types of standard buffer systems have been developed to prevent erosion, protect surface and underground water quality, and drain standing water from crop fields. Buffer systems are by necessity site specific and often require the input of engineers or conservation planners such as the USDA Natural Resources Conservation Service (NRCS) in their design. In fact, the NRCS recognizes many different types of buffer systems as formal conservation practices, including Contour Buffer Strips, Filter Strips, Grassed Waterways, Herbaceous Wind Barriers, Streambank and Shoreline Protection, and Vegetative Barriers. Most of these formal Practice Standards rely upon the use of thick grasses and other vegetation to stabilize the soil or increase water infiltration.

Grassed waterways are a buffer designed to channel surface water runoff from fields, slowing the velocity of the runoff and capturing sediment. Seeding a few legumes or tough wildflowers in grassed waterways can increase their value to beneficial insects.

While the primary purpose of a farm buffer system should never be compromised, it can, in some cases, be constructed with specific broadleaf flowering plants to provide functional beneficial insect habitat. It is critical that a qualified engineer or conservation planner be consulted when designing multipurpose buffers. In addition to technical support through the NRCS and the Xerces Society, local Soil and Water Conservation Districts and state departments of natural resources may be able to provide assistance. A few common types of farm buffers are worthy of special attention.

RAIN GARDENS

A rain garden is nothing more than a filter strip scaled down for a home, business, or school landscape. Typical rain gardens consist of a shallow depression located near a rain gutter downspout or adjacent to a parking lot or other impervious surface. Within the depression, wetland edge plants such as sedges, rushes, native grasses, wildflowers, ferns, and shrubs are planted to stabilize the slopes and encourage water infiltration into the soil.

Well-designed rain gardens can significantly reduce rainwater runoff into storm sewers, which can in turn help limit the movement of pollutants into nearby streams, rivers, and lakes. The dense vegetation of a mature rain garden can also help filter pollutants that might otherwise enter groundwater systems.

To enhance rain gardens for beneficial insects, you'll want to include diverse perennial native wildflowers that bloom throughout the growing season, along with some dense native grasses to provide shelter for ground-dwelling insects. If possible, incorporate some additional structure, such as a couple of stones or logs. Signage, including signs available from organizations like the Xerces Society's Pollinator Conservation Program, or the National Wildlife Federation's Backyard Habitat Program, can help define the space and encourage others to create their own beneficial insect habitat.

Contour buffer strips are bands of perennial vegetation established perpendicular to a slope between annual crops. The buffers slow downhill runoff and soil loss from sloping farmland.

Contour Buffer Strips

Contour buffer strips are a specific type of buffer consisting of narrow bands of perennial vegetation that are planted perpendicular to a slope between strips of annual row crops. This type of buffer is designed to prevent rill or gully formation by capturing sediment as it flows downhill, draining water to a stable outlet on one end of the hillside.

The design of contour buffers usually requires a consistent width along its entire length, usually a minimum of 15 feet, or 4.5 m. Often multiple contour buffer strips will run parallel to each other on a single slope. One typical design, for example, calls for a 15-foot-wide (4.5 m) buffer strip for every 150-foot-wide (45.7 m) strip of crop rows. Because of this parallel placement, contour buffers are usually only established on uniform, nonundulating hillsides. Contour buffer strips are not typically used for slopes with more than a 30 percent grade, and they are not usually used in high-rainfall areas.

*White Dutch clover (*Trifolium repens*) is a low-growing flower that can be easily interseeded into established sod-forming grasses.*

Typical contour buffer strip specifications call for the use of sod-forming grasses; a mixture of sod-forming grasses and legumes, such as clover; or legumes alone. Where legumes alone are used, many conservation engineers recommend doubling the width of the buffer. While nonnative plants are usually selected for this purpose, native sod-forming grasses, or a mixture of native sod-forming grasses along with legumes, wildflowers, and bunch grasses may provide the same functionality. Some engineering guidelines call for contour buffers to be maintained in 15-foot (4.5 m) widths when the vegetation density averages at least 50 stems or blades of grass per square foot. For

ANATOMY OF A CONTOUR BUFFER STRIP

Adding a few deep-rooted perennial wildflowers can increase the pest control and wildlife value of contour buffer strips; however, the majority of the planting should remain in grass for maximum soil stabilization. It's also important to protect such areas from insecticide drift or runoff.

A Cropland
B Contour buffer minimum width 15 feet (4–5 m)
C Drainage

vegetation with lower stem density, it is often recommended that contour buffers be at least 30 feet (9 m) wide.

Many species of native plants and forage legumes can provide the desired erosion-control functionality for a contour buffer, while also providing pollen, nectar, and shelter for beneficial insects. Simply interseeding existing sod-grass buffers with tough, low-growing flowers like white Dutch clover (*T. repens*), self-heal (*Prunella vulgaris*), or yarrow (*Achillea millefolium*) can help enhance biodiversity.

There are several key management points to consider if you want to enhance a contour buffer for beneficial insects. Be sure to choose vegetation that can withstand occasional mowing and equipment traffic. Also, protect your contour buffer strips from pesticides as much as possible. For example, turn spray nozzles off when turning around your equipment inside the strip.

While individually they are narrow strips of land, collectively the area of grass waterways on some farms can be substantial.

Grassed Waterways

A grassed waterway is a graded channel of land that is stabilized with permanent vegetation and designed to carry surface water runoff to a suitable location such as a pond, wetland, or drainage basin. The vegetation covering a grassed waterway helps slow runoff velocity and hold on to soil, preventing erosion. Sod-forming grasses are the best vegetation for this purpose because they can be mowed periodically to allow water to pass efficiently down the center of the waterway.

In some cases, however, simply leaving the side margins unmowed will give beneficial insects shelter and allow room for taller native bunch grasses and perennial wildflowers or small shrubs. In some cases, even within the mowed center of a grassed waterway itself, tough, low-growing wildflowers that tolerate occasional mowing, such as self-heal (*Prunella vulgaris*) or lanceleaf coreopsis (*Coreopsis lanceolata*), can be interseeded into sod-forming grasses. Because of the aggressive growth habitat of the grasses, these wildflowers will not dominate a waterway, but can enhance the area's pollen and nectar resources.

Filter strips use thick vegetation to capture runoff before it can enter streams and irrigation ditches. This California filter strip was intentionally enhanced for beneficial insects with native wildflowers, shrubs, and grasses.

Riparian Buffers and Filter Strips

Another very common type of buffer consists of native vegetation maintained along streams, creeks, and rivers. The primary purposes of these riparian buffers are to shade the water, reducing water temperature for fish and other aquatic wildlife; to filter surface runoff from adjacent farmland; to stabilize stream banks, reducing erosion and downstream siltation; and to provide corridors for wildlife.

To achieve these functions, riparian buffers are usually divided into several management zones. In classic riparian buffer design, the zone closest to the stream bank is maintained in fast-growing native trees, and is usually the narrowest zone; common recommendations call for it to be maintained at around 15 feet (4.5 m) wide. A middle zone of native shrubs, at least 20 feet (6 m) wide, is also common. The outermost zone, consisting of native grasses, is usually the widest at 30 feet (9 m) or more.

Each of these distinct plant communities offers opportunities to include pollen and nectar sources for beneficial insects, as well as shelter for reproduction and overwintering. Simple

*Self-heal (*Prunella vulgaris*) is a tough, weedy wildflower that tolerates occasional mowing and equipment traffic, qualities that allow it to grow in drainage ditches and grassed waterways.*

Although it looks like a natural stream, this California irrigation canal is completely manmade. When native grasses are planted along the banks, the soil is stabilized and wildlife can prosper.

*Native wildflowers like this rattlesnake master (*Eryngium yuccifolium*) are readily available in many areas as seed or transplants from native plant nurseries. The design of new filter strips or riparian buffers should include a diverse range of native plant species wherever possible.*

ways to enhance riparian buffers for conservation biocontrol include interplanting high-value species (including wildflowers in the grass zone and shrubs or trees known to provide nectar and pollen for beneficial insects), and maintaining shelter in the form of stumps, snags, dead trees, brush piles, and bunch grasses. In most cases, you should exclude cattle from riparian buffers, and the tree and shrub zones should be periodically managed to maintain their appropriate width and dominant plant types.

Similar to large-scale riparian buffers, smaller **filter strips** are a type of vegetated structure for protecting downslope streams, wetlands, or irrigation ditches from excess surface water and sediment runoff. Unlike riparian buffers which may have room for large trees and shrubs (such as in a natural floodplain area), filter strips may be as narrow as 20 or 30 feet, and are composed primarily of dense rhizomatous grasses (typically with a stem density of at least 50 stems per square foot). Although the primary function of a filter strip should not be compromised, it is sometimes possible to interplant a few rhizomatous wetland edge wildflowers, such as swamp milkweed (*Asclepias incarnata*) or western goldentop (*Euthamia occidentalis*), into the grasses to enhance beneficial insect resources. Similarly, by diversifying the grass species in the filter strip during the initial planting, and by establishing some wetland emergent species such as sedges and rushes at water's edge, filter strips can provide greater value to more species of wildlife.

Water quality protection is particularly important in dry regions like central California. Well-designed buffer strips not only serve that process, but also contribute to pest control, provide pollinator habitat, and make the farm more beautiful.

ANATOMY OF A FILTER STRIP

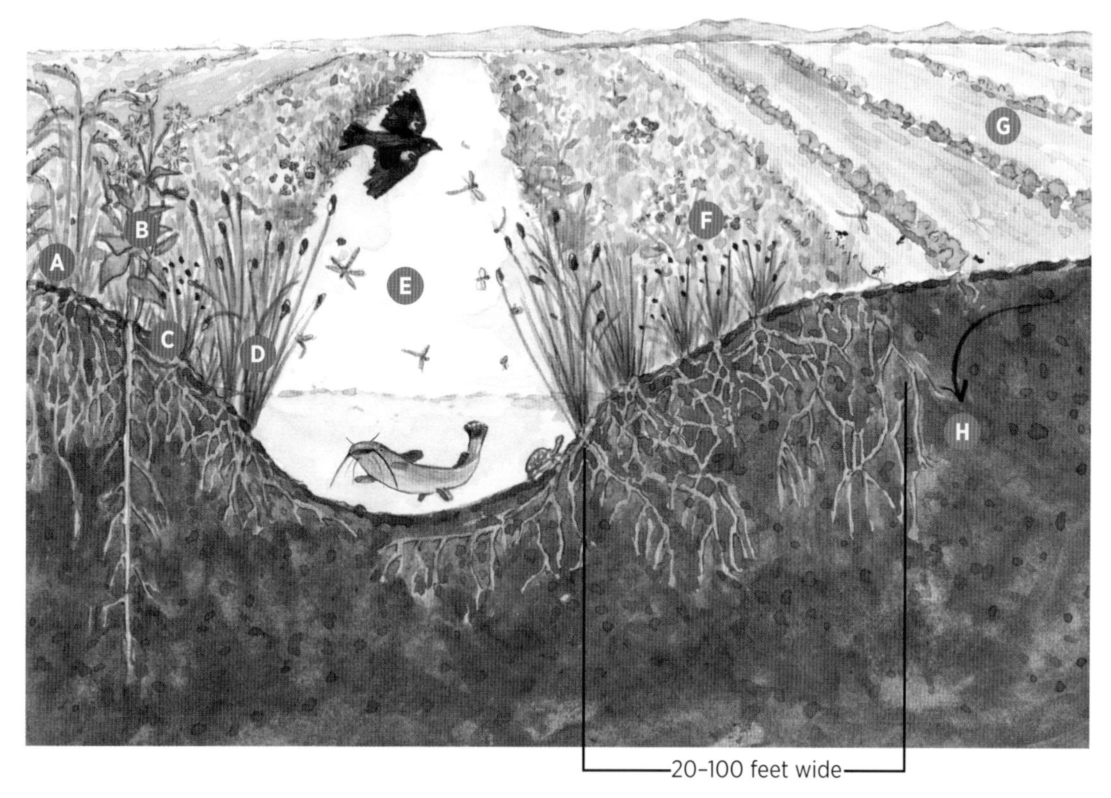

A Rhizomatous native grasses
B Wetland edge wildflowers
C Sedges
D Rushes
E Stream or irrigation ditch
F Sediment capture
G Cropland
H Water infiltration

Minimum of at least 50 stems per square foot

Shelterbelts and Windbreaks

Shelterbelts and windbreaks are linear tree and shrub plantings designed primarily to reduce wind velocity, and conditions related to wind velocity such as soil erosion, snow drifting, airborne odors, and pesticide drift. Secondary benefits of shelterbelts and windbreaks can include visual screening, shade for livestock, harvestable fruit and wood products, wildlife and beneficial insect habitat, and much more.

Most windbreak designs call for 3 to 10 rows of plants, comprising an overall width of 20 to 150 feet (6 to 46 m). Within such a configuration the innermost rows are planted with the tallest tree species, while the outer rows consist of small trees and shrubs. Based upon this design, tall windbreaks are typically able to provide reduced wind velocity for a distance

Most windbreak designs call for 3 to 10 rows of plants, comprising an overall width of 20 to 150 feet (6 to 46 m).

of at least 20 times the windbreak height. For example, a 40-foot-tall windbreak may protect a downwind area up to 800 feet away.

To enhance a traditional windbreak design for beneficial insect habitat, select pollen and nectar-rich shrubs for planting where possible. Farmers can also create additional outer rows that are planted with tall native bunch grasses and wildflowers. In addition, the inner rows of the windbreak can be improved with insect shelters such as brush piles (made up of wood from the windbreak itself) and piles of large field stones.

Conversely, windbreaks can also capture pesticide drift from farm fields, reducing its impact on nearby habitat areas. To provide effective capture of pesticide drift, most design guidelines call for the use of trees with a porous canopy (with 40 to 60 percent open space in the canopy). The idea behind this design is to reduce wind velocity and capture pesticide particles in the windbreak itself, rather than having wind currents simply carry the pesticide particles up and over a denser windbreak where they will be deposited some distance away. Of course, when designing a windbreak for pesticide drift reduction, it is imperative to select plants that are relatively unattractive to beneficial insects, such as conifers.

ANATOMY OF A SHELTERBELT

The tall trees, small trees, shrubs, and wildflowers and grasses incorporated into shelterbelts and windbreaks all contribute to reduction of wind velocity, while providing other benefits to the farm.

100 feet from windbreak, wind speed is reduced to 10 miles per hour

200 feet from windbreak, wind speed is 15 miles per hour

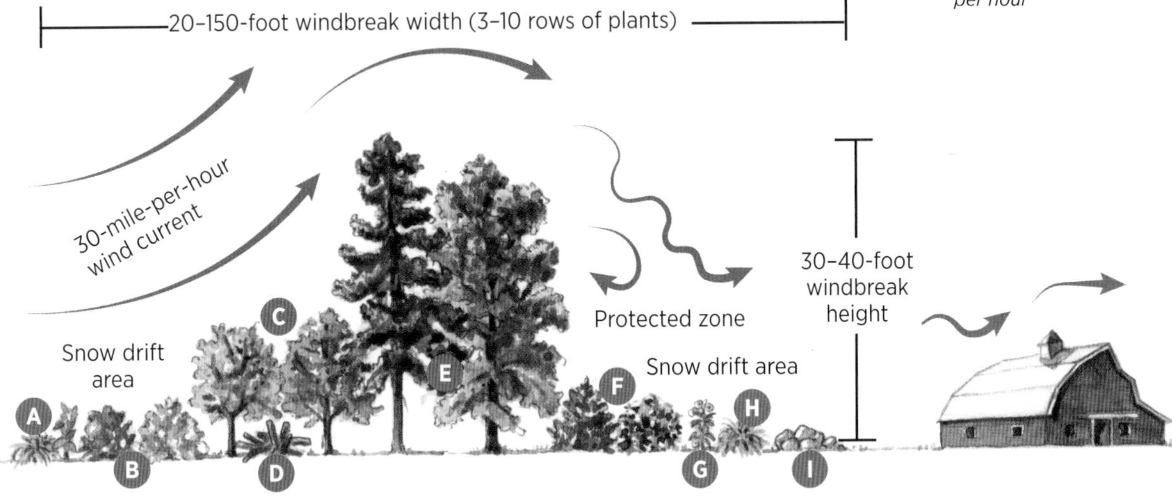

A Rows of native bunch grasses and tall wildflowers
B Shrubs
C Small trees (e.g., hawthorn, wild plum, elderberry)
D Brush piles
E Taller trees (evergreen, or combination of evergreen and deciduous)
F Shrubs (e.g., ceanothus, wild rose, buffaloberry)
G Tall wildflowers (e.g., Maximillian sunflower, showy milkweed)
H Tall native bunch grasses (e.g., big bluestem, deergrass, wiregrass)
I Piles of field stone

COMPARING WINDBREAK DENSITY

40–60% density for wind and pesticide drift reduction

More than 60% density for snow and wind reduction

Organic Farm Buffers

The USDA National Organic Program requirements mandate that organic farmers maintain buffer areas between themselves and neighboring conventional farms as well as other neighboring lands where pesticides might be used, such as roadsides and utility easements. There is no specific size requirement for buffer areas, but they must be sufficient to capture pesticide drift from the neighboring land. An organic certifying agency can provide guidance on the appropriate size for a specific location.

To protect organic crop fields from neighboring conventional fields, some farmers use evergreen windbreaks to reduce insecticide drift.

While organic farmers may grow crops in these buffer areas, they must be harvested and processed separately from other crops on the farm, and they cannot be certified as organic. Consequently, many organic farmers simply maintain buffer areas of grassy fields or weedy tree lines. Although organic buffer areas must be designed to capture occasional agrochemical drift and runoff (and thus are not ideal habitat for beneficial insects), they can be designed and maintained to support as much biodiversity as possible.

For example, an organic farmer might consider placing a few rows of evergreen arborvitae along the property line as an immediate windscreen against pesticide drift, and then plant the inner, protected area of the buffer in native grasses and wildflowers (see the previous section on shelterbelts and windbreaks). Few beneficial insects will use arborvitae for food or shelter, but the trees will help reduce contamination of the inner habitat area.

Similarly, an organic farmer located downhill from a conventional neighbor could maintain the outer edges of the buffer in nonnative sod-forming grasses that support little insect diversity, while an inner strip of the buffer area might be seeded with a more diverse mixture of wildflowers and native grasses. The goal of such a design would be to have the sod-forming grasses capture and filter as much contaminated runoff as possible, while the inner area of the buffer would provide more biodiversity and pest management function.

FOR MORE INFORMATION . . .

A great source for more information about designing and installing farm buffers is the USDA National Agroforestry Center (NAC). See page 241 for more on their outstanding resources, including print publications, website tools, and free software for conservation engineers and farmers.

Buffer strips planted with prairie wildflowers and grasses next to corn and soybeans are a refuge for beneficial insects in Iowa, a state in which row crops dominate the landscape.

CASE STUDY

Buffer Strips for Soybean Aphid Control

THE VALUE OF NATURAL ENEMIES has increased for soybean production with the arrival of the soybean aphid. This invasive pest was discovered in 2000 and quickly spread across the Midwest. Several studies have shown that predators in soybean fields can limit and even prevent soybean aphid outbreaks. The impact of these predators is greatest in landscapes that have more perennial habitat. Natural enemies, especially predators, require alternative prey and shelter from adverse conditions to maintain populations that can feed on crop pests. Annual crops, like soybeans, lack these features, resulting in predator populations that are reduced to a level that is not a significant source of mortality for pests. This is especially true for a state like Iowa, where the landscape is dominated by annual crop production. Adding or restoring perennial habitat may increase the abundance and diversity of predators, setting the stage for more biocontrol.

The landscape of Iowa was once dominated by perennial prairie habitat. At Iowa State University a team of researchers are exploring how adding prairie within a landscape committed to corn and soybean production can increase the delivery of ecosystem services like improved soil and nutrient retention, and increased biodiversity. This project is called the Science-based Trials of Rowcrops Integrated with Prairies, or STRIPs, in part because this targeted approach to conservation results in strips of small amounts of prairie within a crop field.

In our role, we investigated if prairies are a source of aphid-eating predators. To do this, after prairies were established, we surveyed for insect predators in both the prairie and adjacent soybean fields. Over the course of the growing season we observed approximately twice as many aphid-eating predators in prairies than in the soybean field. This suggests that prairies may act as a refuge for beneficial insects, especially during months like May, when soybean — and soybean pests — have not yet emerged.

Initially, at our research site we did not observe more biocontrol in fields with prairie strips compared to fields without the strips. Rather, high rates of biocontrol actually occurred in all of the research fields. As it turns out, we think this similarity may be due to the fact that our study was conducted on fields located within the Neal Smith Wildlife Refuge, the largest reconstructed tallgrass prairie in North America. Thus, the soybean fields embedded within the refuge exist against a background landscape that is not typical for Iowa. To test this possibility, going forward, we are exploring whether the contribution of prairie strips to biocontrol will be greater when habitat is placed in crop fields surrounded by few natural areas.

For more information about the STRIPs project, visit our website at www.nrem.iastate.edu/STRIPs/.

— DR. MATT O'NEAL, Iowa State University, Department of Entomology

Beetle Banks and Other Shelters

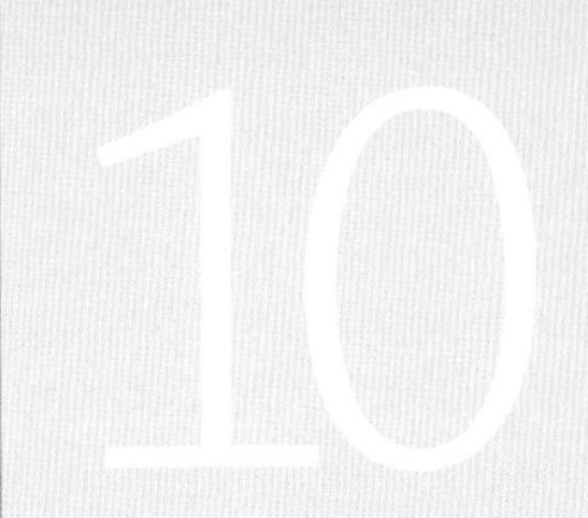

PROVIDING SHELTER for beneficial insects can enhance their ability to survive overwintering, and offers them protected areas for egg-laying and pupation. Common types of shelter include the crowns and understories of tall bunch grasses, thick piles of dead brush, decomposing logs, stumps and snags, and similar features.

- Like other wildlife, beneficial insects need shelter to overwinter and reproduce.
- Where existing shelter is limited, you can create more with simple strategies.
- Diverse types of shelter will support more types of beneficial insects.
- While initial research on shelter types like beetle banks is promising, insect shelter systems remain a relatively new concept.

The exact shelter requirements for most beneficial insects are largely unknown, and little specific guidance for enhancing shelter exists. Often the best advice is simply to leave diverse, uncultivated habitat in place where it already exists. Beyond that, intentionally constructed piles of brush, logs, field stones, and other natural materials are likely to help increase beneficial insect populations.

For a few types of beneficial insect groups, we can offer more specific shelter enhancement strategies. These include beetle banks for predaceous ground beetles, tunnel nest structures for solitary wasps, and brush piles for a variety of beneficial insects.

Beetle banks consist of linear berms of thick grass, the preferred winter cover for predatory ground beetles.

Beetle Banks

Beetle banks are a habitat enhancement intended to provide shelter for a specific group of beneficial insects: predatory ground beetles. Beetle banks consist of long, elevated earthen berms planted with perennial bunch grasses; occasionally the bunch grasses are interplanted with native wildflowers. These banks provide undisturbed winter cover for ground beetles adjacent to cultivated fields, and are intended to promote rapid movement by beetles back into crop fields when warm weather returns the following year.

*While beetle banks are a new concept in North America, farms like this one in Iowa are discovering native grass species such as little bluestem (*Schizachyrium scoparium*) and Indian grass (*Sorghastrum nutans*) that are well suited for the purpose.*

The beetle bank concept originated in Great Britain to provide habitat for beetles and other beneficial insects that had declined due to the loss of hedgerows and other habitat adjacent to cropland. British farmers have used beetle banks successfully to control grain crop pests like aphids and wheat blossom midges; in some cases they have eliminated the need for pesticides altogether. Additional research suggests that various ground beetle species supported by beetle banks may feed extensively on weed seed, and can play an important role in suppressing crop weeds. Despite these promising results, beetle banks remain largely untested in the United States, especially outside the Pacific Northwest.

At maturity, beetle banks are typically thickly vegetated and resistant to weed encroachment.

Planning a Beetle Bank

You can construct a beetle bank by plowing two reverse furrows side by side to create an embankment roughly 2–6 feet (60 cm to 1.8 m) wide and at least 1 foot (30 cm) high. Seed the bank with native bunch grasses or plant it with grass plugs. In western states, blue wild rye (*Elymus glaucus*), California oatgrass (*Danthonia californica*), slender wheatgrass (*Elymus trachycaulus*), and Roemer's fescue (*Festuca idahoensis roemerii*) have been successfully used in beetle bank construction. In other regions native bunch grasses like switchgrass (*Panicum virgatum*), big bluestem (*Andropogon gerardii*), little bluestem (*Schizachyrium scoparium*), Indian grass (*Sorghastrum nutans*), Junegrass (*Koeleria macrantha*), Canada wild rye (*Elymus canadensis*), and prairie dropseed (*Sporobolus heterolepis*) are good potential candidates.

Beetle banks are often positioned in the center of crop fields and may extend almost to the field edges, leaving enough room on each end for equipment to pass. In this way, you can continue to cultivate the entire field around the beetle bank. If your crop field is large, you may need multiple beetle banks positioned at regular intervals to account for the dispersal distance of ground beetles and other beneficial insects. Current guidelines in Britain recommend at least one beetle bank for every 20 hectares (roughly one beetle bank for every 50 acres).

If you want to construct a beetle bank, you should anticipate having to mow it routinely during the first year after planting to suppress annual weeds that may shade and compete with the newly established bunch grasses. In future years you may have to control woody plants and spot-spray or spot-pull weeds. Over time, as the grasses mature, they should be fairly effective at preventing weed encroachment. As with all types of beneficial insect habitat, it is important to protect beetle banks from insecticide spraying.

Beetle banks are still a new concept in the United States, but they closely resemble a number of soil conservation structures already in use by American farmers, including NRCS Practice Standards like Cross Wind Trap Strips or Vegetative Barriers.

EFFECTIVENESS OF BEETLE BANKS

Researchers in Britain wanted to find out whether beetle banks could help reduce cereal aphids in winter wheat. To answer this question, they constructed barriers at various distances from the beetle banks that blocked beetle movement into wheat fields. They then measured the abundance of aphids and aphid predators on either side of the barriers. The researchers found that aphid populations were larger in wheat when barriers blocked predators such as ground beetles, rove beetles, and linyphiid spiders. The impact of the predation by beneficial insects and spiders on aphids declined as distance from the beetle banks increased.

HOW TO

CREATE A BEETLE BANK

Because the beetle bank concept is still very new in North America, there is still very little research on questions such as how big the banks should be. This makes it difficult to provide precise guidance on how best to construct a new beetle bank. Most farmers who are experimenting with beetle banks use one of two common construction approaches: opposite plowing to create a berm, or using a bed shaper.

OPPOSITE PLOWING. To create a berm using opposite plowing, employ a single blade plow, driving carefully in a straight line across the field, pushing the soil to one side in a narrow hill. Lift the plow blade before you reach the end of the field to maintain a vehicle path between the end of the bank and the beginning of the field edge (in the future you can plant crops in this headland area), allowing the entire field to continue to be farmed as a single unit. After creating this initial hill of soil, turn and plow in the opposite direction, pushing soil up against the first hill you created. The result will be a single long mound about 2 feet (60 cm) across.

If you would like a wider bank, repeat the process on both sides, pushing more soil from the sides into the mound. Depending on your soil, the plowing will leave rough clumps that you'll need to smooth before you plant. To do this, drag a lightweight harrow over the hill.

USING A BED SHAPER. A much simpler approach to creating a beetle bank is to use a single-row bed shaper. The disadvantage of this approach is that the width of your beetle bank will be limited to the width of the bed shaper. Despite this, most bed shapers produce a smooth, packed surface, eliminating the need for follow-up harrowing.

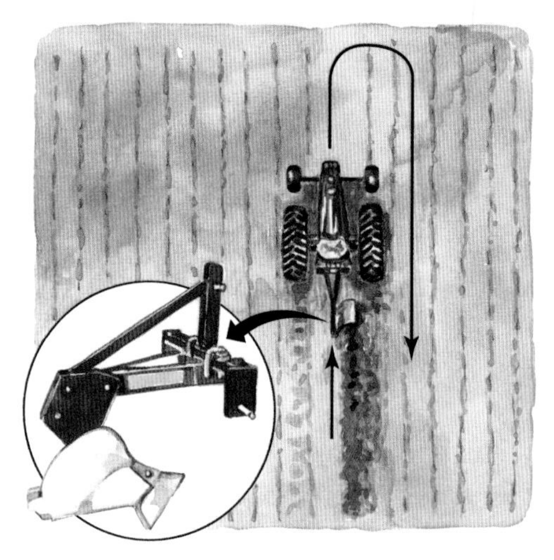

Create the mound for your beetle bank by using a single-blade moldboard plow to build a hill running down the center of your field. By reverse plowing in opposite directions, create a mound roughly 4 feet wide by 2 feet high.

PLANTING A BEETLE BANK

Beetle banks can be established either from seed or with transplants, but they should be planted relatively soon after you've created the mound, to stabilize it. On narrow banks of at least 2 feet (60 cm) wide, you should be able to transplant grass plugs in a staggered pattern around 18 inches (46 cm) apart. Wider banks may require multiple rows of transplants, staggered in a checkerboard pattern to ensure adequate coverage.

Grass seed can also be broadcast by hand over the mound, but expect some of it to show up in the adjacent crop fields. A typical seeding rate for native grass of 5 pounds (2.3 kg) per 1,000 square feet (93 square m) should provide good cover. For even greater beneficial insect value, you might also consider adding a few easy-to-grow wildflowers, as plugs or seed, to the beetle bank. In general, however, your vegetation should consist of at least 75 percent native bunch grasses.

Spring is the best time to establish a new beetle bank; if you begin in the fall, winter precipitation might wash away seed and loose soil. Many native grasses do not require cold stratification to germinate from seed, so spring planting, with a little irrigation as needed, will usually be enough to establish a good cover.

Transplant plugs of native grasses into the mound in multiple rows, spacing the grasses roughly 18–24 inches apart.

BEETLE BANK MANAGEMENT

Cropland weeds will be a problem in most beetle banks during the first year, and sometimes into the second year as well. You should plan to mow the bank if possible once or twice during the first few years, when annual weeds begin to take over. This occasional mowing won't hurt your perennial native grasses, and it can actually help them compete with weeds by encouraging them to send out tillers and expand their crowns.

After several years, the grasses will increase in size and crowd out most weeds. As this happens, you should reduce, and ideally stop, any further mowing since it could harm ground-nesting birds and will definitely reduce winter cover for beneficial insects. In the long term, if shrubby weeds begin to take hold in your beetle bank, you can control them by spot spraying, or pulling them out.

Once grasses are mature, minimize disturbance to the beetle bank so it can provide a stable refuge for beetles and other beneficial insects.

(Continued)

These Oregon farmers created a beetle bank by first reverse plowing to shape the long linear soil mound. A bed shaper, had one been available, would also have worked.

Grass plugs from a native plant nursery were transplanted into the mound, although direct seeding is also an option if plugs are not available.

After establishment the beetle bank integrates into the farm as simply another row in the middle of a vegetable crop field, requiring little maintenance.

Tunnel Nests

Forest edges, woody plants along riparian areas, hedgerows, and snags can all supply nesting habitat for predatory wasps that build nests in wood tunnels. Solitary wasp females nest as individuals, with each female constructing and provisioning her nest, rather than exhibiting the cooperative social interactions that occur in large colonies of paper wasps or hornets. Adult females hunt prey and bring it back to the nest, where they leave it within a cell along with a wasp egg. The wasps construct cell partitions within nests, and cap nest entrances with mud, resin, plant materials, or other materials.

If you're unable to incorporate woody plants into your farm landscape, consider constructing tunnel nests, which many predatory wasps will readily occupy. Tunnel nest structures for solitary wasps are identical to those constructed for solitary wood-nesting bee species such as mason bees and leafcutter bees. In general, you can construct a tunnel nest by drilling a series of dead-end holes into a wood block and hanging it from any available barn, fence post, tree, or other aboveground structure.

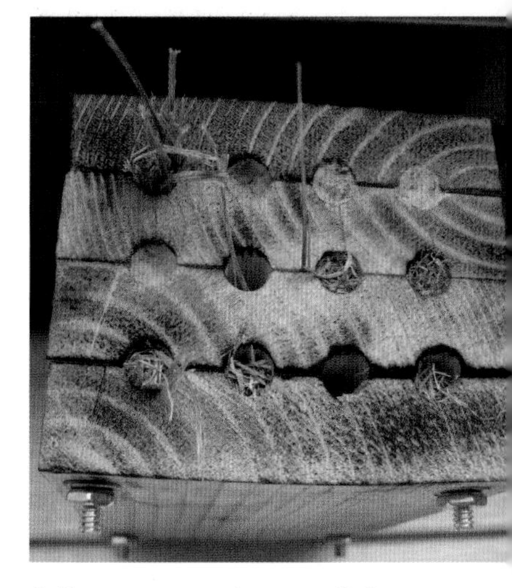

Solitary, nonaggressive wasps that create nests in hollow stems or nest blocks will use mud, grass, or other materials to seal the nest entrance.

Unlike nest blocks designed for solitary bees, tunnel nests for wasps may attract more nesting insects if the holes are relatively small in diameter (1/8 inch or 3 mm for wasps that gather aphids). Offering a range of hole sizes between 1/8 and 5/16 inch (3 and 8 mm) in diameter and 4 to 6 inches (10 to 15 cm) deep, however, will likely attract larger wasp species, as well as solitary bees. Bundles of reeds or pithy stems are also options that will attract predatory wasps. Nests should be sheltered from rain and placed above the height of surrounding vegetation, in a sunny location, and ideally near a visual landmark such as a large tree, or along a hedgerow. Various solitary wasps in the Vespidae and Sphecidae families are common inhabitants of these structures, provisioning their nests with beetle grubs, caterpillars, aphids, and even grasshoppers.

Grass-carrying wasps (Isodontia spp.) provision their nests with grasshopper or cricket prey and use grass in their nest construction.

FINANCIAL ASSISTANCE

Making brush piles and wood block nests for beneficial insects and other wildlife are actions that may be supported by financial incentive programs available through the NRCS (as part of the agency's Upland Wildlife Habitat Management or Fish and Wildlife Structure Practice Standards). Contact your local NRCS office for more information.

Wasp Shelters

In some cases, it may be worth encouraging social wasps to nest near your crop fields. Paper wasps (*Polistes* spp.) are important predators of caterpillars. Researchers in North Carolina found that enhancing wasp nesting habitat near tobacco fields reduced the damage caused by tobacco and tomato hornworms, as paper wasps feed the caterpillars to their larvae.

Paper wasps build nests from chewed plant fibers, creating open cells where brood are reared. The entire nest structure is usually suspended from a stalk anchored in a sheltered area, such as under the eave of a barn. New wasp nests are initiated by a mated queen in the spring; the colony grows throughout the summer and eventually dies, except for a new mated queen, in the fall. By midsummer, paper wasp nests may contain between five and 50 individuals.

CONSIDERATIONS FOR GARDENERS

Habitat features like beetle banks, brush piles, tunnel nests, and insect hotels can all be scaled down for use in garden settings. Recently cultivated garden beds offer little shelter for ground-dwelling beneficial arthropods taking cover from inhospitable weather, predators that would otherwise dwell under leaf litter and mulch, and those beneficial insects that live part of their life beneath the soil. To support these diverse beneficial insects, gardeners can plant small beetle banks between vegetable beds, or even "beetle bumps," small mounds planted with perennial grasses, within vegetable plots.

There are other steps you can take to bolster natural pest control by ground-dwelling arthropods. Using leaf mulch or straw in garden beds can offer shelter to predators such as ground beetles, sheet-weaving spiders, and wolf spiders. Consider providing some nesting opportunities for the predatory wasps that hunt common garden pests such as tomato hornworms and armyworms. You can create small wooden nest blocks with only a few openings or make small stem bundles and hang them throughout your yard for predatory wasps. Consider leaving naturally occurring bare patches of ground for nonaggressive ground-nesting solitary wasps that also hunt garden pests. Finally, insect hotels, creative stacks made from myriad materials, can be ideal for gardeners who like to experiment but are pressed for space.

HOW TO

CREATE SHELTERS FOR SOCIAL WASPS

To create new nesting opportunities for social wasps, some growers construct shelters consisting of wooden boxes that are open on the bottom; old birdhouses can be reused for this purpose. To make your own, cut five 4-by-4 inch (10-by-10 cm) lengths of 1-inch (2.5 cm) board. (Larger dimensions can also be used.) Assemble the pieces to construct a box that is open on the bottom. Attach the nest shelter to a post or tree at least 3 to 4 feet (1 to 1.2 m) off the ground along fencerows or field borders. Wasp shelters should be located away from places with lots of human or animal activity where the nests might be a hazard. Although you can observe the nests if you approach them carefully and slowly, disturbances such as vibrations caused by mowing nearby will disturb them. If disturbed, social wasps will defend their nest by stinging.

CREATE NESTS FOR SOLITARY WASPS

WOODEN NEST BLOCKS

- You can construct wooden nest blocks by drilling holes in various kinds of wood. Blocks of 4-by-4-inch (10 by 10 cm) or 4-by-6-inch (10 by 5 cm) preservative-free lumber are most commonly used, but rougher blocks of wood can also be used. You can drill holes in a stump, an existing snag, or a fence post. Pithy sticks measuring ¾ by ¾ by 6 inches (2 by 2 by 15 cm) may also be drilled and then bundled together.
- On one side of your block, drill a series of holes separated from each other and from the edges by ½ to ¾ inch (1.3 to 2 cm). Holes should be between ⅛ inch (3 mm) and ½ inch (1.3 cm) in diameter. Drill holes as deeply as you can into the block. Holes should be a minimum of 3 inches (8 cm) deep, ideally around 5 to 6 inches (13 to 15 cm) deep, but should not go all the way through the block.
- Cavity-nesting wasps do not like to nest in tunnels open at both ends, which are harder to defend from intruders, so don't drill the block all the way through. Use sharp drill bits, such as brad point bits, to make the smoothest, straightest holes, to increase the likelihood of wasps using the tunnel.

STEM BUNDLES

Sections of bamboo or reeds, such as the common reed (*Phragmites australis*), can also be used as nests. Use plants with hollow stems that are divided into sections by nodes, since these nodes will serve as the closed end of the tunnel nest. You can buy bamboo poles or stakes from garden centers at minimal cost, or collect dead reed stems from wetlands or marshes. You can also use cup-plant stems (*Silphium perfoliatum*) or teasel (*Dipsacus fullonum*).

Cut bamboo and reeds with hand clippers, pruning shears, or a fine-toothed saw. Cut reeds to lengths with one node each, so that each length has one open end only. Use pieces with thick walls and with lengths that

(Continued)

CREATE NESTS FOR SOLITARY WASPS (CONTINUED)

exceed 3 inches (8 cm), but discard pieces that have openings larger than ½ inch (1.3 cm) and smaller than ⅛ inch (3 mm) in diameter, as they are too large or too small to be used by most predatory wasps. Bundle the reeds together, with the closed ends of the stems at one end of the bundle, and secure tightly with string or wire.

Another option is to pack the node ends of the stems into a jar or tin can, so that the open ends face outward. Place the bundles with the stems parallel to the ground. If the bundles contain irregular lengths and diameters, they can shelter various species of wasps, which often have specific preferences for nest cavity dimensions.

NEST PLACEMENT

Hang your nest block with the entrance holes in a horizontal position facing east or southeast, so that morning sunlight reaches the entrance holes early, but isn't too direct during the hottest portion of the day, which can be too harsh for developing brood. Provide protection from rain either by attaching a small overhanging roof to the nest, or by hanging the nest in a protected location such as next to a building.

Nest blocks or bundles should be placed at least 3 feet (1 m) off the ground, and in a visible location, such as on a fence post, or near a large tree, in order to help wasps locate nests in the landscape. Nests may be more rapidly colonized when they are near permanent strips of habitat, such as hedgerows or field borders.

MAINTENANCE

Tunnel nests do need attention over time to avoid the buildup of parasites and diseases that can affect wasp populations. There are several approaches to maintaining nest blocks. Paper liners can be used only with wooden nest blocks and are the most intensive management approach. First, line the tunnels of wooden nest blocks with removable paper straws, which can be purchased or made using waxed paper curled around a dowel rod. At the end of the growing season, gently pull out the straws containing the wasps and place the straws in a cool place, such as a refrigerator or garage. Then place new straws inside the block. Place old straws in an emergence chamber, a dark container such as a plastic bucket with a lid that has a single ⅜-inch (9 mm) hole at the bottom of the side of the container. In the spring, place the emergence container next to the empty block. As wasps emerge from their nests, they will crawl toward and then exit the hole in the emergence container and find the clean nest nearby. Leave the old straws in the emergence box for the season.

If you don't use paper liners, it is important to phase out and replace nesting blocks or bundles regularly. Every two or three years, place entire bundles or blocks in emergence chambers. Place the emergence chambers next to new nests, and the wasps will colonize the new, clean nests rather than returning to the old nests.

The least intensive option is to use multiple smaller bundles of reeds or small blocks with fewer tunnels. Spread the blocks out across your property, which mimics natural conditions of limited nest sources separated spatially. Smaller nests also decompose more rapidly. Add small nests to the landscape periodically, and allow the old nests to deteriorate naturally.

Brush Piles

Intentionally constructed piles of brush, logs, or field stones will provide shelter and overwintering habitat for some beneficial insects. For example, predatory insects like fireflies and lacewings that are active at night may take shelter in brush piles during daylight. Similarly, some solitary predatory wasps will use brush piles as nesting habitat, laying eggs in hollow stems or under exfoliating bark and provisioning those nests with prey for their young. And jumping spiders will spin a silken retreat in which they will overwinter nestled within brush piles.

Brush piles are often considered an eyesore, but the nesting and overwintering shelter they provide for wildlife and beneficial insects is worth recognizing.

BUILD A BRUSH PILE

To create a brush pile optimized for beneficial insects, begin by placing larger branches or logs on the bottom, along with any stones, so that the pile has a sturdy base.

Continue to stack branches and logs, layering them crosswise with small limbs and brush on the top, or pile branches in a conical shape around a large central log. As the brush pile settles, you can add to it or create a new pile.

Good locations for brush piles include along field borders, hedgerows, wooded edges, or fence corners. Several smaller but well-placed brush piles are more beneficial than a single large pile. Create two piles per acre, or more for areas with little natural cover.

Orchard prunings, weedy felled trees, field stones, and even old pallets can supply the raw materials for an insectary brush pile.

If the brush pile is an eyesore, consider highlighting its role in supporting pollinators, pest management, and larger wildlife with education signage for farm visitors.

Insect Hotels

A kind of cross between brush piles and tunnel nests, insect hotels are another way to incorporate artificial shelter for insects into farm landscapes. Insect hotels, like beetle banks, originated in Europe, and are currently popular with gardeners and city dwellers. The value of insect hotels to agricultural landscapes has not yet been thoroughly investigated, but insect hotels may have value in landscapes that are particularly lacking in seminatural, messy areas where insects can overwinter.

Insect hotels can be constructed from a variety of reclaimed materials such as old wooden pallets, used bricks, flowerpots, roof tiles, corrugated cardboard, straw, old leaves and pinecones, sticks and brush, and so on. The idea is to build a structure with or stack these materials to provide insects with nooks and crevices in which they can shelter. Insect hotels could, in theory, provide shelter for lady beetles, lacewings, spiders, predatory wasps, rove beetles, and a number of other beneficial insects, including pollinating bees.

Insect hotels like this one combine the features of brush piles and tunnel nests to attract many kinds of insects.

HOW TO

BUILD A HOTEL

Construct insect hotels on level ground in sun or partial shade. Build your base by stacking and arranging bricks in the shape of an H. If you would like to make a wide insect hotel, arrange two or more H shapes with the bricks, side by side or with gaps in between. Lay a wooden pallet or pieces of plywood across the bricks. Construct a new level using more bricks, clay roof tiles, or clay flowerpots, also in the shape of an H, and then stack another pallet or strips of wood across the top. Continue to stack in this manner until your hotel is as tall as you would you like. You can keep your hotel dry by topping it with roof tiles or a plywood sheet covered in roofing felt. Some people have added a layer of compost and soil to the roofs of their hotels, in order to grow drought-tolerant plants such as sedum.

Once the stack is complete, fill the gaps and spaces between bricks and wood with wood chips, leaf litter, hay, straw, pinecones, rolls of corrugated cardboard, bamboo stems, logs with holes drilled in them, stones, and other materials you might have available. Your creativity can run wild when constructing insect hotels; they can be as fancy or as simple as you choose. Although insect hotels are not a substitute for high-quality habitat, they are an alternative way to provide shelter when it is not otherwise possible.

The diversity of materials used in constructing insect hotels will directly influence the diversity of insects attracted to it. To maximize your impact, use as many types of material as possible — wood, bamboo, stone, brick, straw, and more. Remember also to protect the structure from rain.

CASE STUDY

Banking on Beetles in Oregon

PREDACEOUS GROUND BEETLES are a diverse group with feeding habits that range from specialists to generalists. Some, including the slug- and snail-eating carabid beetle (*Scaphinotus marginatus*), have specialized body parts that enable them to prey specifically on snails. Others, like the ubiquitous genus *Pterostichus,* will eat a variety of other insects or weed seeds. Some species are seasoned travelers, traversing acres of farm fields with ease on their foraging excursions, while others are restricted to small portions of the field, field boundaries, or noncropped habitat. Many fly and are nocturnal, preferring invertebrate nightlife over sunny daytime activities, while others are flightless and can be day-active.

Despite this diversity, one thing all predaceous ground beetles have in common is where they dwell: in the organic duff layer under plants, on or just beneath the soil surface. They are also all highly susceptible to broad-spectrum pesticide exposure and to disturbance of the soil by cultivation practices.

Because predaceous ground beetles are known to consume weed seeds, as well as a diverse variety of crop pests including aphids, cabbage root maggot, Colorado potato beetle larvae, codling moth larvae, snails, slugs, and various fly maggots, researchers and farmers worldwide are experimenting with approaches that increase beetle populations on farms. The most effective method of conserving and enhancing on-farm predaceous ground beetle populations is to provide sheltered areas on the farm for the beetles to colonize. For example, farmers in China have successfully used rounds of fencing placed in the fields and filled with straw to provide temporary refuge for spiders and predaceous ground beetles. At the end of the season, farmers pick up the fencing and till the straw into the soil. In Britain in the 1990s farmers first successfully implemented beetle banks, semipermanent refuges for beetles within their farm fields.

Farmers in Oregon are now experimenting with beetle banks, and developing their own versions planted with native bunch grasses. They selected native bunch grasses because they do not spread invasively on the farm and because the dry, insulating clump of organic matter that forms at their base gives shelter to

Oregon State University was an early champion of beetle banks, an idea that is now catching on across North America.

ground beetles. To create the bank, farmers till a strip within or close to the production field and raise it to about 1 foot (30 cm) above the surrounding field by plowing and reverse plowing. Then they smooth the top and prepare a fine seedbed for the grasses.

At Gathering Together Farm in Philomath, Oregon, two upland prairie grasses, blue wild rye (*Elymus glaucus*) and slender wheatgrass (*Elymus trachycaulus*), and one lowland species, meadow foxtail (*Alopecurus pratensis*), were transplanted onto a bank in spring 2006. A few miles east, at Persephone Farm in Lebanon, Oregon, several beetle banks were sown with these same three grass species. All of Persephone's banks were prepared similarly to the one at Gathering Together Farm, but the grasses were direct-seeded into the bank in the fall. Several of the Oregon beetle banks are over four years old and are still providing excellent cover. Meadow foxtail died off after the second year on all of these banks, but proved to be an easy-to-establish first stage in the planting. Farmers may use this approach to introduce other, slower-growing native bunch grasses, once the bank is established.

Oregon farmers have learned that the more time they spend preparing a weed-free bank site, the less time they spend managing weeds on the banks in the years to come. On a well-established bank, the primary weed management practice used is simply mowing the bank once a year.

Studying these beetle banks in Oregon, researchers have discovered that diverse communities of 33 different beetle species can be found in the adjacent fields. They also have found that populations of predaceous ground beetles are relatively more abundant than at other sites. In addition to good overwintering habitat for beetles, the banks also provide seasonal habitat for spiders and are a great source of of pollen and alternative insect prey, for other beneficial insects such as lady beetles, soldier beetles, and hover flies.

— GWENDOLYN ELLEN, Project Coordinator, Farmscaping for Beneficials, Integrated Plant Protection Center, Oregon State University

PART 3

Managing Beneficial Insect Habitat

WHETHER YOU HAVE RECENTLY installed your beneficial insect habitat or it is already in place, pesticide management and habitat maintenance over time are critical to supporting beneficial insects. Even the best habitat cannot compensate for the exposure of beneficial insects to insecticides. Fortunately, you can take small steps to reduce overall pesticide use and to practice risk-reduction strategies.

Beneficial insect habitat needs management to keep it healthy and productive, but the very management practices that are used can have a negative impact on beneficial insects themselves. It can be a balancing act to introduce disturbance to your habitat to maintain it over time while causing the least amount of harm to beneficial insects. With care and planning, you can sustain your habitat and the ecosystem services it provides for many years to come.

Beneficial insect habitat, including this remnant prairie on an Oklahoma ranch, requires protection from pesticides and long-term land management.

11 Reducing Pesticide Impacts

- Insecticides can harm pests and beneficials alike.
- If pesticides must be used, follow strategies to minimize their harm.
- Even natural or organic pesticides can pose a risk to beneficial insects.
- Alternatives to pesticides are readily available for some pest situations.

PROTECTING AND PROVIDING HABITAT for beneficial insects is the first step in conservation biocontrol. Unless beneficial insects are also protected from insecticides, however, their populations may never increase enough to make a significant contribution to pest control.

The concept of Integrated Pest Management (IPM) provides a decision-making framework for reducing pesticide use in both conventional and organic farming systems. IPM employs a four-phase strategy:

- Reduce conditions that favor an increase in pest populations
- Establish economic thresholds for specific pests (i.e., understand how much damage you can tolerate before you must take responsive action)
- Monitor pest populations
- Apply nonchemical or chemical pest control only when a preestablished damage threshold is reached

The goal of IPM is to reduce pesticide use while still controlling pest populations below economically damaging levels. Conservation biocontrol can be a core component of the first phase of IPM, in which a farmer works to reduce conditions that favor pest populations.

When truly adopted, IPM can help protect beneficial insects, the environment, and you, your family and your farmworkers from unnecessary pesticide use. Cooperative Extension staff can assist with the development of farm-specific IPM programs, and financial support for transitioning to IPM is available through the USDA Natural Resources Conservation Service (NRCS).

Organic farming practices are typically much safer for beneficial insects. Keep in mind, however, that even some organic-approved insecticides can harm beneficial insects.

If You Must Use Pesticides

Insecticides have an impact on beneficial insect populations, whether it's because the poisons kill the insects directly or because they persist in the landscape, causing sublethal effects that inhibit foraging and reproduction among beneficial species. Even herbicides, which do not directly kill insects, can kill native grasses and wildflowers that host beneficial insects and reduce the amount of foraging and egg-laying resources available.

If you must use insecticides, try to apply them only when beneficial insects are not active. Many beneficial insects prefer warm daylight hours for active feeding, so nighttime spraying with active ingredients that have short residual toxicities is a simple strategy for reducing harm. Note, however, that residual toxicity of many insecticides can last longer in cool temperatures, and dewy nights may cause an insecticide to remain active on the foliage the following morning. Also note that some very important pest predators, including most predaceous ground beetles, are nocturnal.

Pesticide Selection

Even when you're careful, insecticides will affect beneficial insect populations. When insecticide use is unavoidable, use products with the lowest possible impact on beneficial insects.

Insecticide labels may list some of the known hazards to beneficial insects. If this information is absent, toxicity information about honey bees, if available, can help you identify how relatively toxic a particular product may be to other beneficial insects. However, some products that are not toxic to honey bees may be toxic to beneficial insects, given that many beneficial insects are much smaller than honey bees and are affected by lower doses of pesticides. Also, unlike honey bees, predatory and parasitoid insects forage wherever their prey populations reside, including where no flowers are present. If they continue to feed on prey dying from pesticide exposure, they or their offspring may die as well.

The best approach to insecticide selection is to choose products, when possible, that are toxic to only a narrow range of insect species. For example, *Bacillus thuringiensis* (*Bt*) is a natural bacterium with many strains, each of which is toxic to a specific group of insects; *Bacillus thuringiensis* var. *kurstaki* (*Btk*) typically is used to control various moth caterpillars. While such selective insecticides may not prevent all harm to nontarget insects, they are a better choice than chemicals with broad-spectrum toxicity to all insect groups. For more information on *Bt*, see Microbial Insecticides and Nematodes (next page).

A newer class of insecticides, known as neonicotinoids, has been promoted in recent years despite being toxic to a wide range of good and bad insects. Because of their low toxicity to mammals, these chemicals have been widely embraced by various agricultural and landscape industries. Neonicotinoid

THE ENVIRONMENTAL LIFE SPAN OF NEONICOTINOIDS

New research is demonstrating that under some circumstances neonicotinoid insecticides are extremely persistent in the environment, with residues sometimes remaining active for years after they were originally applied. Additionally, neonicotinoid concentrations may increase to very high levels in perennial plants that are treated year after year, although how widespread the phenomenon is remains unknown. Finally, there are concerns about flowering weeds absorbing neonicotinoids from the soil around nearby treated plants and potentially poisoning flower visitors like pollinators and other beneficial insects.

Because of these concerns, and because neonicotinoids are highly mobile in soil and water, we recommend caution if they must be used. For more information about the impact of neonicotinoids on beneficial insects, check out the Xerces publication *Beyond the Birds and the Bees: Effects of Neonicotinoid Insecticides on Agriculturally Important Beneficial Insects*, available at: www.xerces.org/pesticides.

products mimic the toxins found in nicotine, and are applied as seed and root treatments, foliar sprays, and trunk injections. The chemicals are then absorbed and transported by the vascular system throughout the plant. Research demonstrates that these chemicals are sequestered in flower nectar and pollen, and nectar- and pollen-feeding insects such as lacewings, parasitoid wasps, and lady beetles may be poisoned as a result. Predatory beneficial insects can also be exposed and harmed when they ingest pests that were exposed but not killed. It is unclear whether neonicotinoids can be integrated with biocontrol. Because of the long-term impacts of these products on beneficial insect populations, we recommend avoiding their use when possible.

Finally, the popularity of organic farming has been increasing at a tremendous rate. However, it is important to note that even organic pesticides can harm beneficial insects. Some organic-approved products are just as lethal as conventional insecticides. Pyrethrin and spinosad, two common pesticides used in organic farming, are broad-spectrum insect killers, destroying pest and beneficial species alike. Some other organic-approved products are safer to use as long as they are not applied where beneficial species are active. Those less-toxic pesticides include horticultural oils and insecticidal soaps.

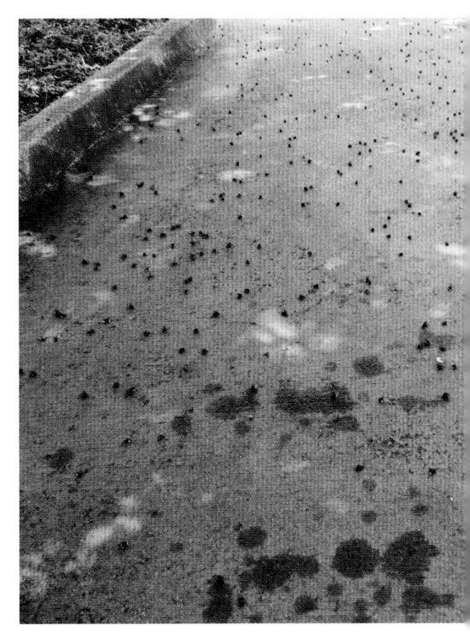

If pesticide labels list a risk to bees, they should be considered potentially harmful to all beneficial insects. The dead bumble bees in this photo were killed when they were exposed to a neonicotinoid insecticide present in the nectar of blooming trees.

Microbial Insecticides and Nematodes

As an alternative to synthetic insecticides, some farmers use microbial insecticides consisting of bacteria, fungi, viruses, or even nematodes for pest control. Most of these products are allowed under organic certification standards, and they are often touted as being safer than conventional insecticides for beneficial insects.

In some cases, this is true. For example, most formulations of the naturally occurring soil bacterium *Bacillus thurengensis* (*Bt*) are selectively toxic to different groups of insects; the bacterium creates holes in the gut wall of insects that feed on it, releasing those gut contents into the bloodstream. Most commercial *Bt* products consist of a specific subspecies, such as *B. thurengensis* var. *kurstaki,* which only affects caterpillar pests, or *B. thurengensis* var. *san diego,* which only affects beetle larvae.

Sprayer equipment strongly influences the potential for pesticide drift. In the case of boom sprayers, they should be operated as close to the crop canopy as possible.

While these toxins may also kill nonpests such as butterfly caterpillars or beneficial beetle species, they are at least considered safe for parasitoid wasps, lacewings, and bees.

In other cases, microbial toxins are less safe for beneficial insects. Spinosad, the active ingredient derived from another soil bacterium, *Saccharopolyspora spinosa,* is toxic to a wide range of insects, including bees and parasitic wasps, when they contact the spray droplets. Spinosad functions by disrupting the nervous system of insects, leading to muscle loss, eventual paralysis, and death.

Similarly, the soil-dwelling fungus *Beauvaria bassiana* also has variable toxicity to many different insect groups, and its effects on some beneficial insects are not well documented. *Beauvaria* is applied as a spray containing the fungal spores. When an insect walks across treated surfaces, it picks up the spores on its body, where they later germinate and produce threadlike hyphae that penetrate the insect's body and result in a widespread infection, eventually killing the insect. Because it is a living fungus, *Beauvaria* tends to require specific environmental conditions (usually cool, humid environments) to thrive. Until more is known about the impact of different formulations on various beneficial insects, use *Beauvaria* with caution.

Insect viruses are available for the control of some pests such as codling moth, corn earworm, and tobacco budworm. The insecticides formulated from these viruses are promoted as being host specific, and are not documented to harm beneficial insects. However, viruses that infect other animal species are sometimes known to undergo genetic recombination that alters their virulence or host range. Because the commercial availability of insect viruses is now fairly limited, the risks to beneficial insects are probably very low. As a broad category, however, the development of new insect viruses should be approached with extreme caution to prevent unintended harm to nonpest insects.

Close-up on Nematodes

Live nematodes (microscopic roundworms) are another group of infectious agents used for pest control. While there are more than 16,000 known parasitic nematodes found in soil and water, only a few species, primarily *Heterorhabditis bacteriophora, Steinernema carpocapsae,* and *Steinernema feltiae,* have been identified and commercially propagated for pest control. These nematodes tend to dwell in the upper soil profile, and are parasites of insects such as beetles and flies, often in the larval stage. Nematodes attack their hosts by sampling the air around them for the presence of insect respiration and thus locating them in the soil. Once a host insect is located, the nematode usually enters it through the mouth, anus, or a tracheal breathing tube and begins feeding, reproducing, and defecating. The latter is typically what kills the host: the host's bloodstream is poisoned with bacteria.

Nematodes are usually commercially available combined with vermiculite for mixing into potting media, or absorbed into sponges for soaking into a tank of water. The water is then subsequently sprayed over crop fields.

While nematodes may help control a wide range of crop pests, including black vine weevils, cucumber beetles, carrot rust flies, squash vine borers, Japanese beetles, and corn rootworms, their wide host range means they are likely to attack many beneficial soil-dwelling insect species as well.

Because of this, the ultimate pest control effectiveness of nematodes may be hard to assess under real-world conditions. It may be that in field crops, released nematodes reduce pest and beneficial insect populations equally and consequently have negligible pest control value. Depending on local conditions, it is even possible that some of the beneficial insects killed by released nematodes may be suppressing secondary pests that are not the target of original nematode release. For example, if nematodes are released into a field for the control of cucumber beetles, and in the process they also reduce populations of beneficial soldier beetle larvae, then aphids, which the soldier beetles were previously preying upon, may explode in numbers.

Because of such interactions, and the uncertainty of how nematodes impact resident beneficial insect populations, we do not recommend their release. It is worth noting, however, that healthy soils with an abundance of organic matter that are protected from compaction typically have existing populations of nematodes, including parasitic species, and may already be providing natural suppression of some pests.

Alternatives to Pesticides

Some farms do not use pesticides at all to maintain crop quality and yield. As unlikely as pesticide-free farming may seem today, consider that it was the norm in the not-so-distant past. In the past, farmers used nonchemical strategies such as:

- Crop rotation to disrupt pest life cycles
- Crop diversity to limit the size of pest populations and ensure against complete loss
- Classic breeding and seed selection for pest-resistant crop varieties

These three fundamental strategies are still used today by farmers who don't use pesticides.

In addition, protective barriers such as floating row covers or spray-on kaolin clay emulsions now provide alternatives to pesticides. For example, apple and pear growers in Japan protect their crops with specialized cloth bags that surround each developing fruit. Amazingly, the majority of Japan's commercial apple crop is produced this way, and special Japanese fruit bags, as well as other similar products, are becoming increasingly available in the United States.

Pheromone traps and mating disruption pheromones are nontoxic pest control strategies that target specific species and complement conservation biocontrol.

Kaolin clay is increasingly used as an alternative to insecticides. Applied as a liquid slurry, the clay covers leaf surfaces to inhibit feeding and egg-laying by crop pests.

Row covers can exclude fruit and vegetable pests and eliminate the need for insecticides in some cases. Keep in mind, however, that they also exclude beneficial insects.

Beyond simple pest barriers, mate-finding pheromones are commercially available for some pest insects and can be sprayed near crop fields or incorporated into dispensers that are hung in crop plants. These pheromones confuse or disrupt the mating of insect pests, making it more difficult for mates to find each other. These same pheromones are often commercially available in the form of monitoring traps that attract and catch pest insects on sticky cardboard landing pads. Other, nonpheromone sticky traps capture pest insects using scents and brightly colored surfaces to which they are naturally attracted (such as red sphere traps used to catch apple maggot flies). Although the primary purpose of these traps is to monitor pest populations, they may also contribute to pest control on smaller farms.

The USDA Natural Resources Conservation Service offers an Integrated Pest Management Conservation Practice, geared toward guiding farmers to adopt pest management practices that have reduced risk to natural resources. Growers interested in IPM can receive cost-share assistance, as well as technical assistance, with the implementation of IPM practices. Through this conservation practice, the NRCS may be able to help farmers work with crop consultants to develop a custom IPM plan for their farm. To learn more, visit your local USDA Service Center (http://offices.sc.egov.usda.gov/locator/app).

Crop-specific guidance on these pesticide alternatives also may be available from your local Cooperative Extension service. When these nonchemical strategies are combined with habitat enhancement for beneficial insects, the result can be farm systems with fewer pest problems.

CONSIDERATIONS FOR GARDENERS

Although you might expect that most pesticides are applied on farms, in actuality more pesticides are applied per acre in suburban settings than in agricultural areas, according to the United States Geological Survey. In 2004, the Environmental Protection Agency estimated that over 78 million homes applied garden or lawn pesticides. Beneficial insects are susceptible to many products used to kill garden or landscape pests, so reducing or eliminating pesticide use is important if you want beneficial insects for pest control.

Controlling Spray Drift

Whenever you apply insecticides, control spray drift to prevent poisoning of beneficial insects and other wildlife in noncrop areas. Spray drift occurs when spray droplets, pesticide vapors, or windborne contaminated soil particles are carried on air currents beyond the crop field. In some cases pesticide drift may be limited only to adjacent field border areas, but even that can pose a problem if you are trying to maintain those areas as beneficial insect habitat. In more extreme cases, pesticide drift has been known to cause damage to more than a mile from the site of application. The weather, your application method, and your equipment settings can all affect the extent of drift.

Weather-related pesticide drift increases under several conditions:

- With greater wind velocity
- Under higher temperatures accompanied by strong convection air currents
- During temperature inversions, when dead calm air is trapped close to the ground

In the case of windborne drift, the effects can be reduced by spraying in the early morning or evening when winds are calmer. Pesticide labels sometimes provide specific guidelines on acceptable wind velocities for a particular product.

For drift caused by warm temperature conditions, midday spraying is less desirable because as the ground warms, rising air can lift the spray particles in vertical convection currents. These droplets may remain aloft for some time and can travel many miles.

Newer spray technology such as tower sprayers and electrostatic sprayers can reduce the potential for insecticide drift.

Drift can also occur during temperature inversions, when spray droplets become trapped and move laterally above the ground in a cool, lower air mass. Inversions often occur when cool night temperatures follow high day temperatures; they are often characterized by foggy conditions.

Optimal spray conditions for reducing drift occur when the air is slightly unstable, with a very mild, steady wind of 2 to 9 miles (3 to 15 km) per hour. Ideally, temperatures and humidity should be moderate. Contact your local Cooperative Extension for specific guidance about your region.

Your spray application methods and equipment settings also strongly influence the potential for drift. Since small droplets are most likely to drift long distances, avoid aerial applications and air-blast sprayers whenever possible. Operate standard boom sprayers at the lowest effective pressure and set the nozzles as low to the ground as possible. Nozzle type also has a great influence on the amount of drift a sprayer produces. Select nozzles capable of operating at low pressures (15 to 30 psi) to produce larger, heavier droplets that will deliver

PESTICIDE PROTECTION RESOURCES

There are several web resources that can help you choose pesticides that are least harmful to beneficial insects.

The University of California system has a statewide Integrated Pest Management program to dispense advice. Their main website provides information about IPM and crop-specific guidance about pesticides and alternatives for managing pests. The crop index can guide you to pages specific to the crop of your choice, and provides a link to pages with information about the relative toxicities to beneficial insects of pesticides used in that crop. You can find information about pesticide selectivity or mode of action, and the duration of their impact to beneficial insects in the crop index at: www.ipm.ucdavis.edu/PMG/crops-agriculture.html.

The publication *How to Reduce Bee Poisoning from Pesticides*, from Pacific Northwest Extension, focuses on ways to protect bees from hazards associated with pesticides, and contains tables that provide the toxicity of various pesticides to bees. While many pesticide labels may not list hazards to other beneficial insects, products with labels that mention a hazard to bees should also be considered toxic to all beneficial insects. A PDF of the publication is available here: www.xerces.org/pesticides.

the insecticide within the crop canopy, where it is less likely to be carried by wind currents.

New electrostatic sprayers can also help reduce off-target pesticide applications. These sprayers apply the pesticide with special nozzles that electrically charge the droplets, which are then attracted to the leaf surfaces. This approach delivers chemicals more effectively and efficiently than traditional nozzle technology does, and can reduce off-target applications by over 50 percent. Regardless of the chemical or the type of application equipment you use, make sure your sprayers are properly calibrated to ensure that you're not applying excess amounts of pesticide.

Finally, nonflowering windbreaks and conservation buffers can act as barriers to reduce pesticide drift from neighboring fields. For example, windbreaks of dense evergreen trees, which typically attract relatively few beneficial insect species, can be used as a simple barrier for reducing pesticide drift and protecting adjacent beneficial insect habitat. The USDA Natural Resources Conservation Service can provide guidance and financial support for the construction of pesticide drift barriers.

Air-blast sprayers and aerial crop dusting create pesticide mist that can be difficult to keep precisely on target.

CASE STUDY

Designing Windbreaks to Limit Pesticide Drift

FOR FARMERS WANTING TO PROTECT insect habitat from pesticide drift, or protect their organic crops from a neighbor's conventional pesticide drift, a carefully designed windbreak can be an effective tool. Trees and shrubs, particularly small-needled evergreens, are known to be exceptionally good at capturing spray drift. Don't use plants that attract beneficial insects, of course.

Research shows that a windbreak that allows some of the wind to pass through it (a feature described as **porosity**) is more effective than one that doesn't. A solid windbreak of overly dense trees deflects the wind upward, creating eddies on the leeward side that could bring drifting pesticides back down to the surface, an effect known as **downwash**.

The best pesticide drift protection comes from a windbreak made of several rows of trees and shrubs that include small-needled evergreens. These trees are two to four times as effective as broadleaf plants in capturing spray droplets, and provide year-round protection. While a windbreak with multiple rows of trees is the optimum, even a single row can substantially reduce drift if space is limited.

The shape, structure, and width of the windbreak can all affect its droplet-capture effectiveness. The ideal windbreak is made up of five rows of small-needled evergreens, starting with a shrub row on the windward side, with the rows behind increasing in height. Minimum height at maturity should be one and a half times the spray release height.

Spacing between rows should be 12 to 20 feet (3.6 to 6 m) and should allow room for mowers and other equipment. The exact distance will be guided by the mature width of the plants and your own equipment and maintenance practices. Where possible, row spacing should be closest on the windward and leeward sides, and farthest between the innermost rows. Designs with a mixture of shrubs and trees or with fewer rows can be planted a little more densely.

The reality is that this ideal may not be achievable. Where not enough space is available to establish five rows of trees, even just two rows of evergreens can help reduce drift. However, if the

The best windbreaks for capturing pesticide drift are evergreens that allow some of the wind to pass through them, while at the same time capturing most of the drift on their needles.

windbreak is not dense enough, it may need to be twice the spray height to provide meaningful benefits.

Generally, you should align windbreaks to intercept prevailing winds, although the location will be dictated by site conditions and available space. While some crops benefit from being sheltered from wind, others may not thrive with less light, so your design needs to balance wind reduction with the effects of shade. You might place the windbreak on the windward side of a field or farm to protect it from drift, or you might want to place it instead on the leeward side of crop fields to prevent movement of chemicals off-site and into adjacent insect habitat.

When designed correctly, windbreaks can be effective, but they make up only one component of best management practices for minimizing chemical drift. Other key actions include timing your spraying to avoid high wind conditions and active times for beneficial insects; selecting appropriate nozzles, with the understanding that smaller droplets travel farther and are less easily captured by vegetation; and regular sprayer calibration to avoid overapplication.

Windbreaks provide a unique opportunity to address conservation threats to beneficial insects and, at the same time, address a wide variety of other resource concerns, from crop production and reduced soil erosion to wildlife habitat. They continue to be a flexible and useful tool for conservation on agricultural lands and an important feature for sustainable farms.

— NANCY LEE ADAMSON, THOMAS WARD, MACE VAUGHAN,
The Xerces Society and USDA Natural Resources Conservation Service
Adapted from *Inside Agroforestry*, Vol. 20.

CASE STUDY

Biological Mite Control in Pennsylvania Apple Orchards

AT THE FOOT of the Appalachian Mountains, in the main fruit-growing region of southern Pennsylvania, lies a three-generation family farm that produces fresh market apples and tart processing cherries. Now run by three grandsons of the founder, the 800-acre Lerew Brothers Farm in York Springs was among the first in the state to move away from broad-spectrum organophosphate and carbamate insecticides to control leafroller and codling moths. In the mid-1990s the Lerews transitioned to using more pest-selective insecticides that were also safer to farmworkers.

Throughout this transition, and in the years since, the Lerews worked closely with the nearby Pennsylvania State University (PSU) Fruit Research and Extension Center to adopt models for predicting pest outbreaks that guide the timing of insecticides. As a result, the Lerews were able to reduce the number of sprays for the main moth pests from seven per season to only three. They also enjoyed much-improved fruit quality.

Later, in 2003, PSU entomologist Dr. David Biddinger began to find greater numbers of beneficial insects in the Lerews' orchards. These good bugs significantly reduced outbreaks of secondary pests such as mites, aphids, and leafminers. Most striking was the lack of outbreaks by the European red mite, a major pest in the past. Most other fruit growers at the time were averaging $150 to $200 per acre in pesticide costs, with about one-third of this being spent on mites. Because mite predators survived in the Lerews' orchards, they had not sprayed for pest mites in more than five years.

In typical Pennsylvania orchards, PSU researchers found only three insecticide-resistant mite predators, a *Stethorus* lady beetle and two species of predatory mites. These predators were only partially effective in reducing mite injury. In the Lerews' orchards, however, pest mite populations were 10 to 15 times lower than in other growers' orchards; mite damage to the leaves never occurred. We discovered that the predatory mite *Typhlodromus pyri*, which had never before been seen in Pennsylvania, was responsible. This mite had been used in effective biocontrol programs in New York, New England, and Washington state orchards, but never in the mid-Atlantic. It had been generally thought that *T. pyri* was a cool-weather predator that could not adapt to warmer regions. An intensive survey of other Pennsylvania farms in 2004 found *T. pyri* to be present at low levels on farms that were using selective insecticides, but this beneficial mite was not seen at all on farms still using broad-spectrum insecticides.

The majority of predators move into the orchard only when pest populations are high, and already causing significant leaf damage. In contrast, *T. pyri* spends its entire life on the tree, feeding on pollen and fungal spores at times when the pest mite populations are very low. Because it never leaves the apple tree, *T. pyri* is a much more effective mite predator and could effectively regulate pest mite populations throughout the season. However, because *T. pyri* lives permanently on the apple trees, a toxic insecticide application anytime during the season would completely eliminate it.

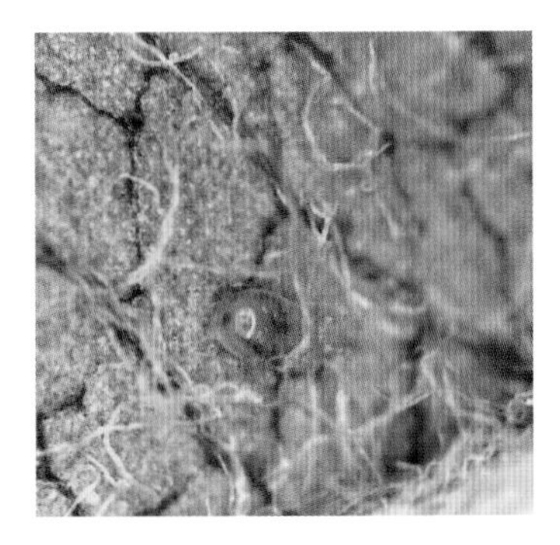

Although no bigger than the period at the end of this sentence, predatory mites can help keep spider mites and rust mites in check in fruit and vegetable crops.

Typhlodromus pyri can be easily transferred to new orchards by clipping branches and placing them in other orchards. Using the Lerew Brothers Farm and the PSU research orchards as two *T. pyri* "seed sites," PSU quickly established the predator in the majority of Pennsylvania's 22,000 acres of apples. This expansion was facilitated by financial incentives provided to the growers through NRCS conservation programs, and informed by conservation guidelines provided by PSU. Current Penn State University recommendations for establishing and conserving biological mite control with *T. pyri* are regularly updated at: http://extension.psu.edu/ipm/resources/nrcs/programs/conventreefruit/biocontrolmites/view.

Because of this conservation biocontrol program, it is estimated that the amount of miticides used in Pennsylvania orchards has been reduced by more than a ton of active ingredients each year, and growers save more than $1 million in pesticide costs.

— DR. DAVID BIDDINGER, Biocontrol Specialist & Senior Research Associate, Penn State University Department of Entomology;
JIM LEREW, Farmer, Lerew Brothers Farm;
DR. ED RAJOTTE, Professor of Entomology and IPM Coordinator, Penn State University Department of Entomology

12

Long-Term Habitat Management

- The best habitats for beneficial insects tend to be sunny, open areas, which can require maintenance.
- Even well-intentioned land management practices can adversely affect conservation biocontrol.
- Balance land management with beneficial insect ecology by limiting and rotating disturbance.
- Properly done, land management can enhance the value of insect habitat.

WHEREVER BENEFICIAL insect habitat exists, it is important to think about how you can maintain it over the long term. Depending on your circumstances, there might be times when habitat conservation must be balanced with the need to control weeds adjacent to cropland, or to manage activities like grazing.

The best beneficial insect habitat typically consists of open, sunny spaces such as meadows, prairies, or shrubby areas. This is fortunate because each of those habitats typically tolerates occasional disturbance. Mature forests may provide beneficial insect habitat on their edges, but most forest insects have different ecological requirements from those of the common beneficial insects found on farms. Similarly, grassland areas may not have as many beneficial insects if they lack wildflowers and are dominated by only a few species of grasses; thus occasional disturbance can actually be helpful by temporarily reducing dominance by some grasses, and encouraging the germination of dormant wildflower seed.

With these thoughts in mind, to conserve areas of the farm as beneficial insect habitat you should prioritize the protection of sunny areas with diverse mixes of native grasses and wildflowers. Large trees that may shade out these native plants should not be allowed to encroach, and grasses, especially nonnative grasses, and other weedy plants should not be allowed to dominate and reduce overall plant diversity.

Even well-designed, high-quality beneficial insect habitat sometimes needs spot weeding and other maintenance.

To achieve this ideal balance, you may need periodically to pull weeds by hand or spot-spray them with herbicides. Think however, about whether the need to remove weeds outweighs their value to beneficial insects. Obviously you should remove noxious weeds, and weeds that support pest insects, but noninvasive flowering weeds can supply essential pollen and nectar food resources and ensure that your predator and parasitoid insect populations remain high even when their prey is absent.

You also might need occasionally to fell large, weedy trees if they encroach on restored habitat, or girdle them and leave them standing for beneficial insects and raptors to nest in, if they don't present a safety hazard. By tolerating natural features like standing or fallen dead trees, stumps, or trimmed brush piles, you are protecting sites that provide shelter to beneficial insects.

Disking, Mowing, and Burning

On larger conservation lands (such as land enrolled in the USDA's Conservation Reserve Program, or CRP), disking, mowing, and burning are common approaches to maintaining diverse grass and wildflower plant communities. If you use these management practices, minimize them as much as possible during the growing season so that insects can use pollen and nectar resources and ground-nesting birds can raise their young.

If mowing must be conducted in non-crop areas of the farm, it is best to perform it as infrequently as possible, and on a rotational basis, leaving some unmowed patches intact. Other recommendations include moving at a slow speed, using a wildlife flushing bar, and mowing outside the nesting season of birds and beneficial insects.

Disking is a somewhat common practice on long-term conservation easement land, such as CRP. Typically, USDA guidelines recommend that large prairie-type habitats be disked every five or six years if the wildflower species begin to decline and the prairie becomes dominated by grasses or encroached upon by tree seedlings. The disturbance caused by both the tractor and a simple disk harrow is often considered enough to set back the growth of grasses, and to create new, light-filled spaces for dormant wildflower seed to germinate and reestablish. It is important to recognize, however, that such disturbance may also stimulate the growth of dormant weed seed if it is present. This is less of a concern for land that has been

Disking is sometimes used in mature grasslands on a rotational basis every few years to suppress grass growth and encourage the germination of dormant wildflower seed. Keep in mind, however, that opening spaces in the ground can also encourage the germination of dormant weed seed. To increase success, we recommend broadcasting additional wildflower seed over newly disked areas when using this management practice.

maintained primarily as a restored native plant community for several years than for land with recent weedy plant populations. Note that disking is only recommended as a management practice on restored lands. Consult a natural resources agency for guidance on how to maintain intact natural areas.

Periodic mowing can prevent woody trees from encroaching. Where mowing is needed, several other strategies can help reduce harm to beneficial insects and other wildlife. These include mowing during the day and using a flushing bar to help move animals before the blades reach them; cutting at speeds of less than 8 miles (13.5 km) per hour; and cutting as high as possible, ideally not less than 12 inches (30 cm).

Managed fire, known as prescribed burning, is the third common practice used to maintain open, sunny habitats like prairies, meadows, and savannahs. The plant communities that make up those ecosystems are often fire-tolerant, so fire can be an effective way to remove nonnative weeds or seedling stage trees. Despite the initial benefits to some native plant communities, the effects of fire on insect populations can be highly variable. In general, fire is fine on a limited scale and an occasional basis. Widespread and intense fires that don't conserve nearby, unburned habitat have been demonstrated

Prescribed burning can be a useful tool for maintaining open prairie and meadow habitats; however, it is important to limit the scale and frequency of such burns to avoid the loss of local insect populations.

to reduce populations of some insects, and in some cases may lead to local extinctions. To avoid this possibility, limit the size of the prescribed burn (see Rotating Habitat Disturbance, next page), leave any small, unburned patches that occur in the fire area as micro-refuges, and avoid very hot fires by burning during a cool season, either late fall or very early spring.

Grazing

Livestock grazing may also be a viable way to manage beneficial insect habitat. Ideal grazing patterns are rotational in nature, sometimes mimicking the grazing patterns of animals like bison, and help maximize native grass and wildflower diversity. There has been a recent resurgence in this type of grazing, with new expert communities developing across the country. If you're considering grazing as a land management practice, consult an experienced practitioner or a farm conservation agency such as the USDA-NRCS.

With thoughtful planning, grazing can be compatible with native plants and beneficial insect habitat in many areas. Such grazing systems, however, often require specialized expertise to design.

Rotating Habitat Disturbance

If you use mowing, disking, burning, or grazing to maintain beneficial insect habitat, try to minimize harmful impacts by dividing the on-farm habitat into separate management zones, none of which should comprise more than 30 percent of the total beneficial habitat. Ideally, only one management zone, or 30 percent or less of the beneficial insect habitat on each farm, should be disturbed in a single year, meaning that each zone has a three- to five-year management rotation. This will allow beneficial insects to recolonize disturbed areas from surrounding, undisturbed habitat. In natural areas, some habitat should be left in long-term refuge for species that do not tolerate disturbance.

Interseeding Wildflowers

One way to help mitigate the disturbance caused by disking, mowing, burning, or grazing, and to improve floral diversity where it has declined, is to interseed newly disturbed areas with additional wildflower seed. For small areas, you can scatter the seed by hand on bare ground; for larger areas you can broadcast seed mechanically. To improve results, try interseeding before rain or snowfall, both of which will help push the seed below any thatch and into the surface of the soil. You can achieve more effective interseeding over large areas by using a specialized no-till native seed drill, which is designed specifically for planting into an existing thatch layer. On a small scale, if you want to save costs on interseeding a meadow area, you can use wildflower seed that you hand collect from existing plantings on your land.

In habitat areas where wildflower abundance is limited, specialized seed drills and other equipment can be used to interseed additional wildflowers into existing vegetation.

Cover crops, like this buckwheat, contribute to soil health in the Kerr Center's vegetable fields and provide nectar for wasps, flower flies, and other insects that help keep pests in check.

CASE STUDY

Beneficial Insect Habitat on an Oklahoma Farm and Ranch

THOUGH OFTEN OVERLOOKED as a resource for smaller wildlife, rangeland and native pastures can be carefully managed to supply diverse and abundant flowering plants, which in turn can support beneficial insects. At its 1,776-acre farm and ranch in southeast Oklahoma, the nonprofit educational foundation Kerr Center for Sustainable Agriculture offers demonstrations of how to integrate economic and ecological goals to achieve successful ranching and farming operations that provide for biodiversity. Managing grazing animals, pasture, and rangeland is important in Oklahoma, where cattle are the number one farm commodity and raising them is common on both small and large farms throughout the state.

The Kerr Center implements **rotational grazing**, also known as controlled grazing, or management intensive grazing, a practice used to maintain soil fertility, improve forage quality, and extend the grazing season on pastureland. Goats, cattle, pigs, and poultry are rotated through pastures at the Kerr Center. The animals graze pastures intensely for brief periods of time, followed by long rest periods. Although some especially palatable wildflowers cannot withstand the pressures of intensified grazing, others are able to recover and bloom during the rest periods.

As a result of multispecies and rotational grazing, preliminary observations indicate that there is increased biodiversity on the ranch and that something is always in bloom during the growing season. This includes native wildflowers like wild indigo

Oklahoma's Kerr Center is demonstrating that agriculture and native ecosystems can coexist and benefit one another.

(*Baptisia* spp.) and ironweed (*Vernonia* spp.), as well as yarrow (*Achillea millefolium*), coreopsis (*Coreopsis* spp.), Mexican hat (*Ratibida columnifera*), and winecup (*Callirhoe* spp.). These flowers all provide pollen or nectar for predatory and parasitic insects. Introduced species such as arrowleaf clover (*Trifolium vesiculosum*) and hairy vetch (*Vicia villosa*) are also attractive to both beneficial insects and pollinators.

In addition to its pastures, the Kerr Center also grows organic horticulture crops, including heirloom vegetables. Many practices used at the Kerr Center support beneficial insects. Alternatives to insecticides, such as kaolin clay, are used to protect heirloom tomatoes from pests and reduce the impacts of pest management on beneficial insects. Organic insecticides like neem and pyrethrum are used as a last resort and are spot-sprayed to limit the exposure of pollinators and other beneficials to otherwise harmful materials. Cover crops such as rye (*Sercale cereale*) and vetch (*Vicia* spp.) are planted for winter cover, and sudangrass (*Sorghum x drummondii*) and cowpeas (*Vigna unguiculata*) are used in the summer. These cover crops shelter and sustain beneficial insects throughout the year.

Insectary plantings of buckwheat (*Fagopyrum esculentum*) and annual sunflowers (*Helianthus annuus*) provide additional pollen and nectar during the summer and early fall. Beetle banks have been planned for the organic row crop areas, and habitat plantings of native perennial wildflowers intended to support predators, parasitoids, and pollinators are currently being established.

— GEORGE KUEPPER, Horticulture Manager, and DAVID REDHAGE, Director of Ranch Operations and Natural Resources, Kerr Center for Sustainable Agriculture

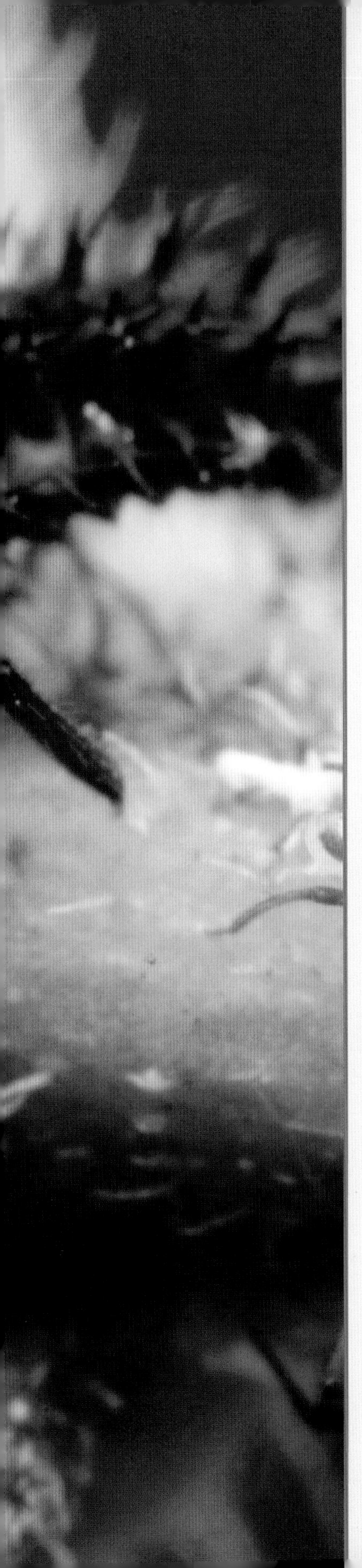

PART 4

Common Beneficial Insects and Their Kin

THIS SECTION WILL HELP YOU recognize important groups of beneficial insects that can be found on farms. You will also find information on their natural history, life cycles, and conservation needs.

Recognizing beneficial insects during their various life stages can be a challenge. Some groups undergo **incomplete metamorphosis**, a gradual, simple transition from egg to nymph to adult (see page 4). Nymphs frequently resemble smaller versions of adults, though they have no wings and cannot reproduce yet. Other groups undergo **complete metamorphosis**, developing from egg to larva to pupa to adult. Insect larvae in these groups often look very different from adults and have different habitat needs. You may readily recognize some groups, while it may take time to become familiar with others.

Lady beetle larvae are adept predators of aphids and other pest insects.

Assassin Bugs, Ambush Bugs

ORDER: Hemiptera

FAMILY: Reduviidae SUBFAMILY: Phymatidae

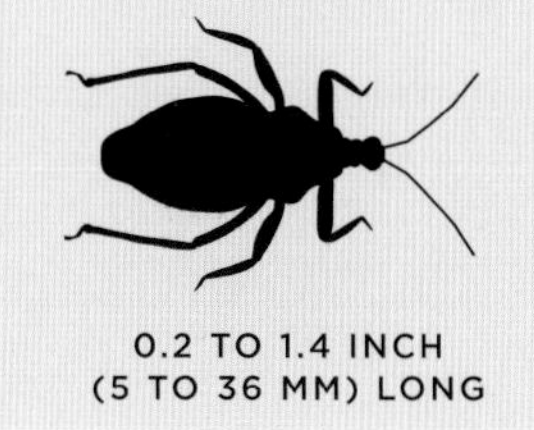

0.2 TO 1.4 INCH (5 TO 36 MM) LONG

Assassin Bug adult

Ambush Bug adult

ADULT ASSASSIN BUGS may be a drab gray or brown, or may have bright colorations. They have an elongate head and a long, slender beak that is used to pierce prey. Many species have swollen, spiny front legs modified for grasping prey (called raptorial legs). Nymphs may somewhat resemble adults but lack wings.

COMMON PREY: Assassin bugs are generalist predators of aphids, grasshoppers, caterpillars, beetles, and various other insects, including other beneficial insects. Assassin bugs are large, aggressive predators, and will often kill more prey than they need for consumption. Wheel bugs (*Arilus cristatus*) are especially valuable predators of caterpillars and Japanese beetles.

SPECIES IN NORTH AMERICA: Approximately 160

DEVELOPMENT TIME (FROM EGG TO ADULT): Four or more weeks

GENERATIONS PER YEAR: One

EGG-LAYING SITES: On leaves or branches of plants

OVERWINTERING: As eggs, nymphs, or adults at the base of plants, under leaf litter, or tree bark

ADDITIONAL HABITAT: Ambush bugs dwell on flowers waiting for prey. They may drink nectar when prey is scarce.

CONSERVATION STRATEGIES: Assassin bugs can benefit from the shelter, overwintering habitat, and alternate prey provided by cover crops, insectary plantings, and permanent plantings such as hedgerows.

Big-Eyed Bugs

ORDER: Hemiptera

FAMILY: Geocoridae GENUS: *Geocoris*

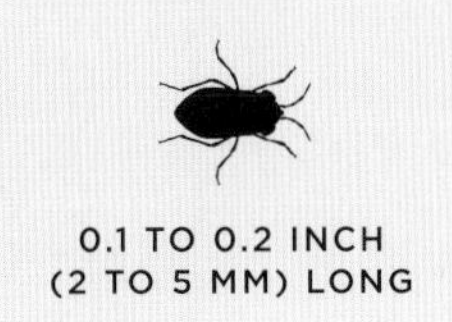

0.1 TO 0.2 INCH
(2 TO 5 MM) LONG

SMALL, OVAL IN SHAPE, adult big-eyed bugs are named for their bulbous eyes that project from the side of the head and help them spot their prey. These bugs are black or pale in color; nymphs resemble adults except that they are smaller and lack wings. Both nymphs and adults are predaceous, with piercing, needlelike mouthparts.

COMMON PREY: Eggs, nymphs or larvae, and small adults of true bugs, beetles, caterpillars, flies, thrips, mites

SPECIES IN NORTH AMERICA: Approximately 25

DEVELOPMENT TIME: Approximately 50 days

GENERATIONS PER YEAR: One to multiple, varies with species

EGG-LAYING SITES: On leaves or in plant litter near prey

OVERWINTERING: Adults overwinter in thatch or under low-growing plants.

ADDITIONAL FOOD SOURCES: Nectar, plant sap, seeds

ADDITIONAL HABITAT: Herbaceous vegetation, leaf litter

CONSERVATION STRATEGIES: Permanent plantings of bunch grasses or shrubby garden plants like oregano or thyme can provide winter shelter to big-eyed bugs. Flowers that provide nectar and seeds can sustain big-eyed bugs when prey is limited. A refuge from tillage is important so that overwintering big-eyed bugs can persist.

Big-Eyed Bug eating Whitefly eggs

Big-Eyed Bug adult

Damsel Bugs

ORDER: Hemiptera

FAMILY: Nabidae

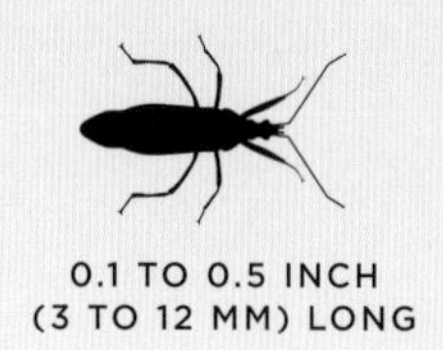

0.1 TO 0.5 INCH (3 TO 12 MM) LONG

Damsel Bug adult

Damsel Bug adult

DAMSEL BUGS ARE SMALL, with a slender body shape and light brown to black coloration. They have needlelike mouthparts, and enlarged raptorial forelegs. Damsel bug nymphs are smaller versions of the adults.

COMMON PREY: Caterpillars, aphids, leafhoppers, leaf beetles, thrips, spider mites, insect eggs

SPECIES IN NORTH AMERICA: Approximately 40

DEVELOPMENT TIME: Five or more weeks

GENERATIONS PER YEAR: Up to five

EGG-LAYING SITES: Eggs are inserted into plant tissue (e.g., leaf blade or petiole).

OVERWINTERING: In fields under debris as adults or as eggs

ADDITIONAL HABITAT: Crop debris, mulch, or brush piles

CONSERVATION STRATEGIES: Provide and protect habitat in which damsel bugs can find sources of alternate prey and places to overwinter, such as meadows, pasture, grasslands, and grassy areas around gardens and near crops. According to research, damsel bugs are more abundant in no-till systems than in tilled systems, and in intercropped systems where two or more crops are grown together than in single species plantings.

Minute Pirate Bugs, Insidious Flower Bugs

ORDER: Hemiptera

FAMILY: Anthocoridae

0.1 TO 0.2 INCH
(2 TO 3 MM) LONG

ADULT MINUTE PIRATE BUGS are tiny, with a flattened, oval-shaped body, a triangular black head, and triangular patterns on the wings. Nymphs are brown or orange in color, have a teardrop-shaped body, and are wingless. Minute pirate bugs have needlelike mouthparts used to puncture prey; both the nymphs and adults are predaceous.

COMMON PREY: Thrips, mites, scales, aphids, plant lice, small caterpillars (including bollworm and corn earworm), and various insect eggs. Pirate bugs can consume around 30 small insects or eggs per day, and excel at seeking out prey at low densities.

SPECIES IN NORTH AMERICA: Approximately 100

DEVELOPMENT TIME: Three weeks or more (slowed by cooler temperatures or lack of prey)

GENERATIONS PER YEAR: Two to three

EGG-LAYING SITES: Eggs are inserted into plant tissue or under bark.

OVERWINTERING: Adults overwinter in leaf litter or under bark.

ADDITIONAL FOOD SOURCES: Pollen, nectar, some plant sap

Minute Pirate Bug adult

ADDITIONAL HABITAT: Leaf litter, herbaceous vegetation, trees

CONSERVATION STRATEGIES: Minute pirate bugs can be supported through the maintenance of permanent plantings within or near to crops, and through protection of natural grassland or wooded areas nearby. Habitat with herbaceous or shrubby plants can provide shelter and alternate food sources. Particularly attractive flowers include plants in the bean, sunflower, and carrot families, as well as willows, elderberry, and buckwheat. Crop diversity can also increase minute pirate bug populations.

Predatory Stink Bugs

ORDER: Hemiptera

FAMILY: Pentatomidae SUBFAMILY: Asopinae

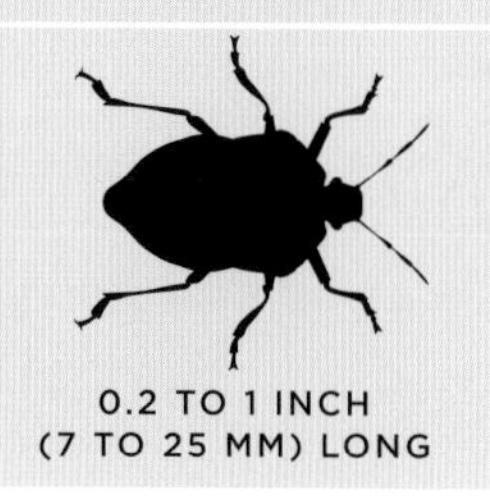

0.2 TO 1 INCH (7 TO 25 MM) LONG

Predatory Stink Bug nymph

Predatory Stink Bug adult

THE COMMON NAME stink bug comes from the unpleasant-smelling substance that can be released from glands on the side of the thorax when the bug is under duress. The eggs of stink bugs are dark and appear barrel-shaped. Although adults clearly have shield-shaped bodies, the bodies of nymphs are often more rounded. Predatory stink bugs have a beak that is very thick at the base, a characteristic that can help distinguish them from plant-feeding stink bugs, whose beaks are slender at the base. Predatory stink bugs may be brown, or they may have bright color patterns.

COMMON PREY: The majority of stink bug species are plant feeders, and some are pests, but those in the subfamily Asopinae are predators. These predatory stink bugs may consume a variety of pest insect eggs, larvae, and small adults, but may also attack small beneficial insects.

SPECIES IN NORTH AMERICA: Approximately 35

DEVELOPMENT TIME: Seven or more weeks

GENERATIONS PER YEAR: 1 or more; varies with species and location

EGG-LAYING SITES: Eggs are laid in clusters on vegetation.

OVERWINTERING: Adults overwinter in leaf litter; some species may overwinter as eggs.

ADDITIONAL FOOD SOURCES: May occasionally feed on nectar and possibly pollen

ADDITIONAL HABITAT: In field crops, meadows, hedgerows, and forest edges

CONSERVATION STRATEGIES: Native plant field borders and hedgerows can provide alternate prey and critical shelter for overwintering eggs or adults.

Mantids

ORDER: Mantodea

FAMILY: Mantidae

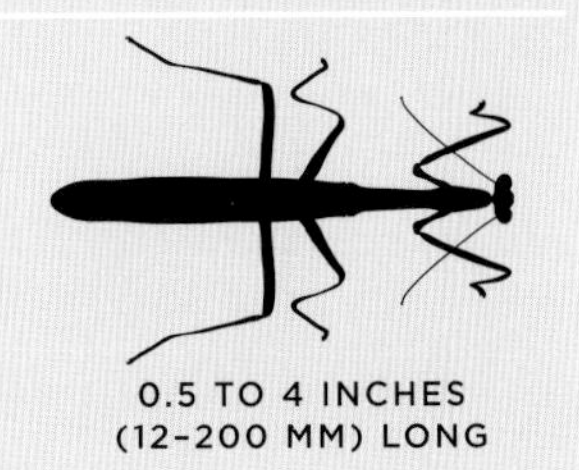
0.5 TO 4 INCHES (12–200 MM) LONG

MANTIDS, SOMETIMES generally referred to as praying mantids, have unique, spiny raised front legs, a triangular head, and an elongated body. Typically green or brown in color, they are well camouflaged in vegetation. Mantids also have excellent eyesight, and can turn their heads back and forth, which aids them in hunting prey. They are not very effective biocontrol agents, however, because they feed indiscriminately.

COMMON PREY: Mantids are nondiscriminating, opportunistic predators of a variety of insects, including aphids, grasshoppers, beetles, bees, and wasps.

SPECIES IN NORTH AMERICA: Approximately 20

DEVELOPMENT TIME: 10 to 12 months

GENERATIONS PER YEAR: One

EGG-LAYING SITES: Eggs are laid in groups on plant stems or branches and are surrounded by a frothy liquid that hardens into a cream or brown protective egg case.

OVERWINTERING: As eggs, enclosed in a protective case attached to plant stems

ADDITIONAL HABITAT: Mantids that reside on flowers produce more eggs and have a higher body mass than mantids that dwell on nonflowering plants.

CONSERVATION STRATEGIES: Native plant buffers can provide alternate prey sources, as well as shelter for overwintering egg cases. Egg cases of the introduced Chinese mantid are available for purchase, but we don't recommend this because it further introduces the species to areas beyond its natural distribution.

Mantid adult

Mantid protective egg case

Green Lacewings, Brown Lacewings

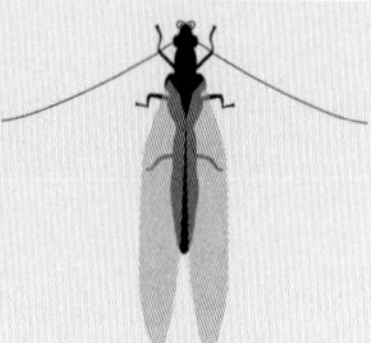

ORDER: Neuroptera

FAMILY: Chrysopidae, Hemerobiidae

GREEN: 0.6 TO 1 INCH (15 TO 25 MM) LONG; BROWN: 0.3 TO 0.5 INCH (6 TO 12 MM) LONG

THE EGGS OF GREEN lacewings are tiny white ovals atop threadlike silk stalks attached to foliage singly or in groups. In contrast, the pale cream or pink eggs of brown lacewings are laid singly on foliage without stalks. The larvae of both green and brown lacewings are similar, with gray-green or brown alligator-like bodies and sharp, long, sickle-shaped jaws. Brown lacewing larvae are slightly more slender than green lacewing larvae and are known to wag their heads from side to side. Lacewings pupate inside silken cocoons attached to the undersides of leaves or in other sheltered sites (e.g., under bark). Green lacewing adults have a pale green body, with eyes that are coppery metallic in color, long, threadlike antennae, and delicate, membranous wings. Adult brown lacewings are very similar to green lacewings, though they are smaller in size and brownish in color.

COMMON PREY: Aphids, small caterpillars, beetles, thrips, mites, whiteflies, mealybugs, and other small, soft-bodied insects. Lacewing larvae are especially voracious predators; they may travel up to 100 feet (30 m) in search of prey on foliage and can consume up to 400 aphids each week, earning their nickname "aphid lions."

Green Lacewing adult

SPECIES IN NORTH AMERICA: Green lacewings: approximately 90; brown lacewings: approximately 60

DEVELOPMENT TIME: Approximately 40 days

GENERATIONS PER YEAR: Multiple

EGG-LAYING SITES: Lacewing eggs are laid on foliage near prey (e.g., near aphid colonies).

OVERWINTERING: Either as prepupae within cocoons attached to leaves, or adults in sheltered areas such as leaf litter

ADDITIONAL FOOD SOURCES: Adults are predaceous or feed on honeydew, nectar, or pollen.

ADDITIONAL HABITAT: Green lacewings are more often seen in crops, fields, and gardens, while brown lacewings are more abundant in forests, orchards, and fields with wooded edges. Lacewing adults are most active at night, and brown lacewings can cope with relatively low temperatures and are active in cooler temperatures than other predators.

CONSERVATION STRATEGIES: In the absence of pollen and nectar provided by flowering plants, adult lacewings may not lay eggs and may disperse elsewhere in search of food. Support lacewings by planting a variety of flowers near crops that have varying bloom times throughout the growing season. Windbreaks or trees near field edges may offer additional habitat. Humidity provided by dense grass may also be important for preventing desiccation of young lacewing larvae.

Checkered Beetles

ORDER: Coleoptera

FAMILY: Cleridae

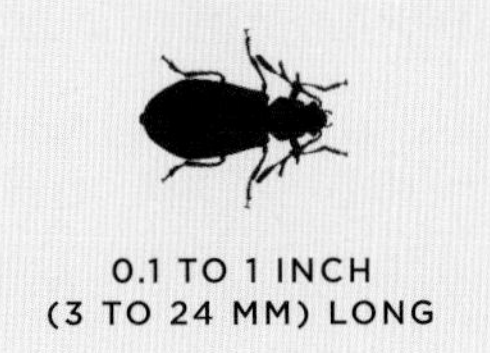

0.1 TO 1 INCH
(3 TO 24 MM) LONG

ADULT CHECKERED BEETLES are frequently brightly colored, with contrasting color patterns on their wing covers. They have elongated, narrow bodies and are covered with soft, short hairs.

COMMON PREY: Most species are predators as both adults and larvae. Some are common on flowers and foliage, feeding on grasshopper eggs, aphids, or other small insects, while many other species are associated with woody plants, feeding on wood-boring beetles or other pests of trees.

SPECIES IN NORTH AMERICA: Approximately 270

DEVELOPMENT TIME: From five weeks to one year or more, depending upon prey availability

GENERATIONS PER YEAR: One to two, depending on species, and prey availability

EGG-LAYING SITES: Under tree bark or in soil

OVERWINTERING: Larvae, pupae, or adults overwinter in soil or under bark

ADDITIONAL FOOD SOURCES: Pollen

ADDITIONAL HABITAT: Adults can be found on bark or flowers. Pupae are found in soil.

CONSERVATION STRATEGIES: Field borders, insectary plantings, hedgerows, and brush piles provide shelter, alternate prey, pollen, and overwintering habitat for checkered beetles.

Checkered Beetle adult

Checkered Beetle adult

Checkered Beetle adult

Firefly Beetles, Fireflies, Lightning Bugs

ORDER: Coleoptera

FAMILY: Lampyridae

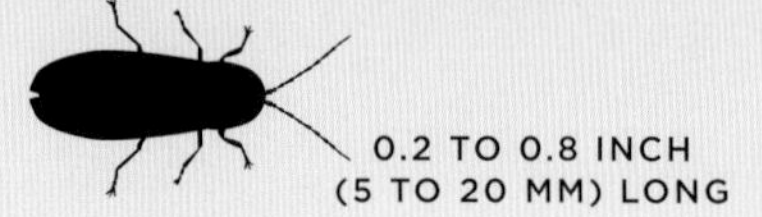

0.2 TO 0.8 INCH (5 TO 20 MM) LONG

Firefly adult

ADULT FIREFLIES have soft, leathery wing covers. They superficially resemble soldier beetles, but most can be distinguished by the light-producing segments near the end of the abdomen. Female fireflies have shorter wings and fewer luminous segments than males, and many species are wingless. The predatory larvae have strong, sicklelike jaws, and are referred to by some as "glowworms" because they are also luminescent.

COMMON PREY: Snails, slugs, caterpillars, and other soft-bodied insects in soil and moist or semiaquatic habitats

SPECIES IN NORTH AMERICA: Approximately 125

DEVELOPMENT TIME: Up to a year

GENERATIONS PER YEAR: One to two

EGG-LAYING SITES: Eggs may be laid singly or in clusters just under soil or among grass roots.

OVERWINTERING: Larvae overwinter under bark or in soil.

ADDITIONAL FOOD SOURCES: Some adults are predatory or feed on nectar and pollen.

ADDITIONAL HABITAT: Larvae reside in damp areas where prey is found, and under bark. Fireflies pupate in soil, under rocks, or in leaf litter.

CONSERVATION STRATEGIES: Tall grass in field edges or nearby habitat can shelter adults and should be protected or supplemented. Reduce tillage to protect egg-laying sites as well as larval habitat and overwintering sites. Flowers with an open structure and exposed nectaries, such as those in the sunflower family, may attract pollen- and nectar-seeking adults.

Ground Beetles

ORDER: Coleoptera

FAMILY: Carabidae

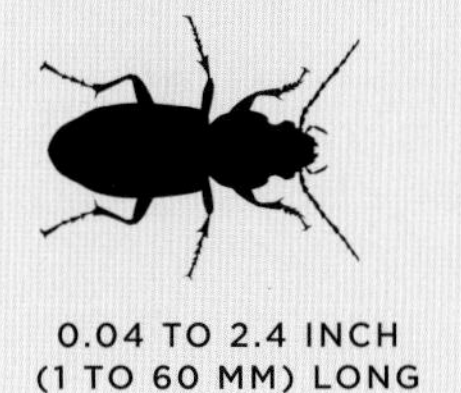

0.04 TO 2.4 INCH
(1 TO 60 MM) LONG

CREAM TO BROWN IN COLOR, ground beetle larvae have a round head with hooked jaws, long legs, and bristly posterior projections. Adult ground beetles range in size from small to large, and have threadlike antennae, prominent eyes, a head narrower than their thorax, an extended-oval abdomen, and ridged wing covers. Their coloration is dark and shiny, usually black or brown, with green, blue or purple iridescence in some species.

COMMON PREY: Insects, including caterpillars (e.g., gypsy moths and tent caterpillars), grasshoppers, beetles (e.g., Colorado potato beetle), aphids, flies, snails, and slugs. Both larvae and adult ground beetles can eat their body weight in prey each day. Most ground beetles feed nocturnally, though some species are active during the day. Adult ground beetles tend to feed on the soil surface or on vegetation, while larvae of predacious ground beetles usually feed under the soil surface on rootworms, caterpillars, and other soft-bodied insects. Larvae are known to kill more prey than they can eat.

SPECIES IN NORTH AMERICA: Approximately 2,500

DEVELOPMENT TIME: One year or more, with adults living up to four years

GENERATIONS PER YEAR: One

EGG-LAYING SITES: Ground beetle eggs are laid singly or in small batches in crevices or, more frequently, in the soil; some females provide protection by laying individual eggs in small chambers in the soil or within cocoons of soil particles; females of some species may guard brood.

Ground Beetle adult

OVERWINTERING: Larvae or adults in grass clumps

ADDITIONAL FOOD SOURCES: Some species of ground beetles are omnivorous and will also eat carrion or fungus, while others feed primarily on seeds of several common weeds.

ADDITIONAL HABITAT: Ground beetles can be found under debris, logs, in soil cracks, or moving along the ground or on vegetation.

CONSERVATION STRATEGIES: Create permanent plantings near crops to support ground beetles. Beetle banks, earthen ridges within fields planted with species of bunch grasses, can provide important overwintering habitat for ground beetles. Mulched areas can also be a refuge, but avoid excessive tillage or burning of crop residue, as these practices can impact multiple life stages of these beetles and can reduce populations quickly. Research also suggests that animal manure and composts benefit the beetles over chemical fertilizer. Weed seed predators are more often found in fields with surface residue rather than bare, fallow fields.

Tiger Beetles

ORDER: Coleoptera

FAMILY: Carabidae SUBFAMILY: Cicindelinae

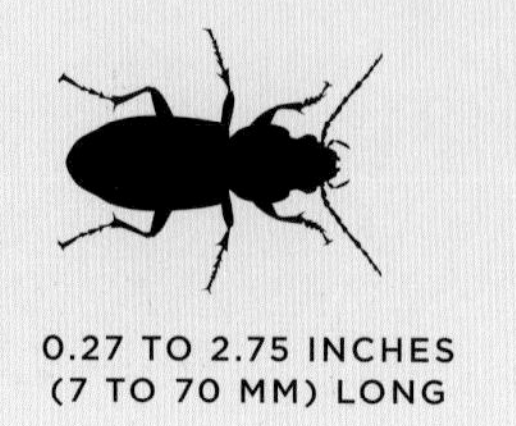

0.27 TO 2.75 INCHES (7 TO 70 MM) LONG

Punctured Tiger Beetle adult

TIGER BEETLE LARVAE anchor their grublike bodies within their burrows and conceal their large jaws under their flat heads while waiting for prey to wander near the burrow. Adult tiger beetles may be grayish brown, metallic bronze, or iridescent blue or green, sometimes with white markings on their wing covers. Adults have long, thin legs used for running swiftly in short bursts. As visual hunters, adult tiger beetles also have prominent eyes, with heads that are wider than the more slender thorax. Large sickle-shaped jaws are used to capture and chew prey.

COMMON PREY: Caterpillars, grasshoppers, beetles, flies, and other insects. Closely related to ground beetles, tiger beetles are less common in many agricultural fields, but may still contribute to pest management. Both larvae and adults are carnivorous. Tiger beetle larvae typically hunt from cylindrical burrows in the soil, grabbing prey that wander by. Adult tiger beetles hunt on the surface of the ground, and are extremely fast-moving. Adult tiger beetles typically hunt during the day, though some are active at dusk or are nocturnal.

SPECIES IN NORTH AMERICA: Approximately 109

DEVELOPMENT TIME: One year or more, with adults living two to three years

GENERATIONS PER YEAR: One

EGG-LAYING SITES: Tiger beetle eggs are laid individually in burrows in damp, well-drained soil.

OVERWINTERING: Larvae or adults within burrows in the soil

ADDITIONAL HABITAT: Tiger beetles favor open ground in grasslands and forests, and can also be found along roadsides or in sandy areas like lake or stream edges and dunes.

CONSERVATION STRATEGIES: Create permanent plantings, like beetle banks or field borders, near crops to support tiger beetles. Maintain open, sandy areas for tiger beetle hunting and nesting; protect stream or lake edges, and reduce tillage.

Six-Spotted Tiger Beetle adult

Tiger Beetle with prey

PREDATORY INSECTS

Lady Beetles, Ladybugs, Ladybird Beetles

ORDER: Coleoptera

FAMILY: Coccinellidae

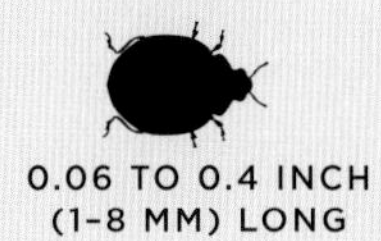

0.06 TO 0.4 INCH (1–8 MM) LONG

LADY BEETLE EGGS are elliptical, yellow-orange in color, and are usually laid in clusters on leaves and stems near prey. Larvae have an elongated, flattened, alligator-like body, and typically dark coloration with bright bands or spots. Adult lady beetles have oval, convex bodies. They are brightly colored with dark markings or black or beige with red or yellow markings.

COMMON PREY: Most lady beetle larvae and adults are specialist predators of aphids or scales, but some also consume whiteflies, mites, thrips, and insect eggs in the absence of their preferred prey. A single lady beetle may consume up to 5,000 aphids in its lifetime.

SPECIES IN NORTH AMERICA: Approximately 475

DEVELOPMENT TIME: Approximately 45 days

GENERATIONS PER YEAR: One to three

EGG-LAYING SITES: On leaves or stems near prey

OVERWINTERING: Adults overwinter in protected locations, such as leaf litter, in rock crevices, behind bark, or in the eaves of homes.

ADDITIONAL FOOD SOURCES: Adults also consume pollen, nectar, and honeydew.

ADDITIONAL HABITAT: Vegetation in agricultural fields, natural areas, and gardens

Convergent Lady Beetle adult

Convergent Lady Beetle larva eating Aphids

CONSERVATION STRATEGIES: Noncropped areas or plantings within fields that support plants producing pollen and nectar at varying times of the growing season can provide adult lady beetles with valuable nonprey food. Flowers that are shallow or open in shape, such as flowers in the sunflower or carrot family, may be most attractive to lady beetles, due to the accessibility of pollen and nectar. Areas with tall or dense grass can provide a welcome humid microclimate for beetles in the summer and potential overwintering sites. Several species of lady beetles have been introduced into the United States for classical biocontrol. Some of these species, such as the multicolored Asian lady beetle and the seven-spotted lady beetle, may be outcompeting and displacing native species.

Soft-Winged Flower Beetles

ORDER: Coleoptera

FAMILY: Melyridae

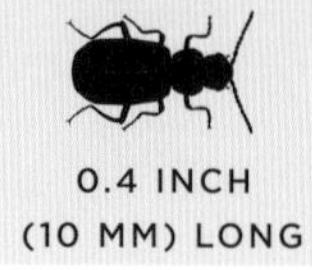

0.4 INCH (10 MM) LONG

Soft-Winged Flower beetle adults

THESE SMALL BEETLES have soft wing covers and an elongate, slightly ovular body that is brightly colored.

COMMON PREY: Adults are frequently found on flowers, where they feed on other insects as well as on pollen. Larvae are primarily predaceous and are common in soil, leaf litter, or under bark. Caterpillars, thrips, stink bug eggs, aphids, and spider mites are all prey for soft-winged flower beetles. Beetles in the genus *Collops* are common predators of pests in cotton, sorghum, and alfalfa.

SPECIES IN NORTH AMERICA: Approximately 520

DEVELOPMENT TIME: Nine weeks or more

GENERATIONS PER YEAR: One to three or more, depending on species and climate

EGG-LAYING SITES: Eggs are laid in clusters in leaf litter, on soil surface, or under bark.

OVERWINTERING: Little is known about overwintering habitat, but it is thought that the beetles overwinter as eggs, larvae, or adults.

ADDITIONAL FOOD SOURCES: Adults also consume pollen and possibly nectar.

ADDITIONAL HABITAT: Eggs of some species are laid in the soil, while others lay eggs on dead plant material. Larvae typically inhabit leaf litter or soil surface.

CONSERVATION STRATEGIES: Cover crops, insectary plantings, field borders, or hedgerows can provide shelter and leaf litter for larvae, and nonprey food for adults. Brush piles and beetle banks may also provide shelter for larvae.

Soldier Beetles

ORDER: Coleoptera

FAMILY: Cantharidae

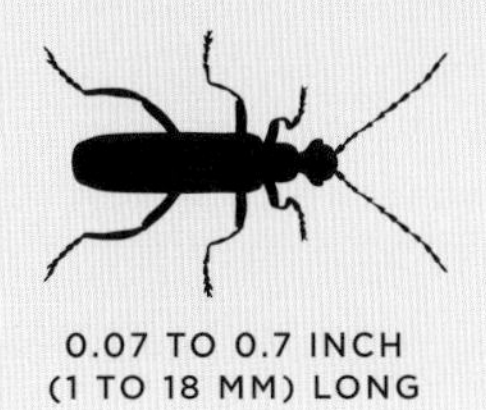

0.07 TO 0.7 INCH (1 TO 18 MM) LONG

ADULT SOLDIER BEETLES are soft-bodied with leathery wing covers. With their black, brown, yellow, or orange coloration and elongated bodies, many resemble fireflies, minus the light-producing abdomen. The dark, flattened larvae hunt for other insects in loose soil, leaf litter, under rocks or debris, or under bark, and some are known to be active hunters even during cold weather.

Soldier Beetle adult

Soldier Beetle larva

COMMON PREY: Insect eggs and larvae, aphids, snails, and slugs

SPECIES IN NORTH AMERICA: Approximately 470

DEVELOPMENT TIME: About a year

EGG-LAYING SITES: Eggs are laid in moist soil or leaf litter.

GENERATIONS PER YEAR: One to two

OVERWINTERING: Larvae overwinter in leaf litter.

ADDITIONAL FOOD SOURCES: Adults also consume pollen and nectar.

ADDITIONAL HABITAT: Pupation occurs just below the soil surface.

CONSERVATION STRATEGIES: Soldier beetle adults need pollen and nectar to survive, and appear to prefer flowers with open structure and accessible pollen and nectar, such as those in the sunflower and carrot families. They are also common on other flower species that have clusters of small flowers or flowers with easily available nectar, such as milkweeds. Protect or supplement habitat near or within fields to support pollen and nectar sources. Protect egg-laying sites as well as larval habitat and overwintering sites in fields by avoiding soil fumigants and insecticides, and reducing tillage. There is some evidence to suggest that larvae prefer areas with higher plant cover and humidity to bare ground, so areas of constant plant cover, such as cover crops or permanent buffer plantings adjacent to crop fields, may benefit larvae and their migration into nearby crops.

Rove Beetles

ORDER: Coleoptera

FAMILY: Staphylinidae

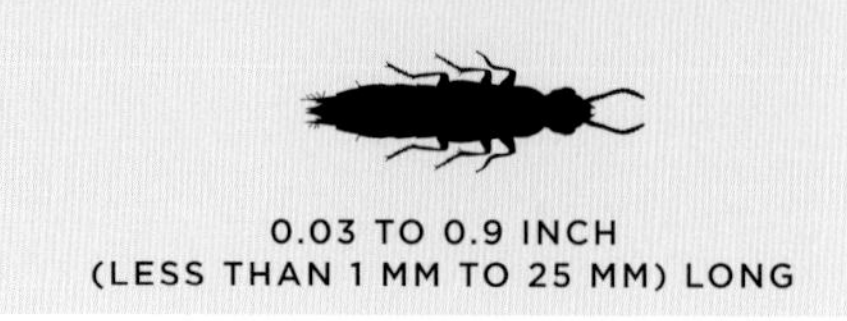

0.03 TO 0.9 INCH
(LESS THAN 1 MM TO 25 MM) LONG

Rove Beetle adults

Rove Beetle larva

ROVE BEETLES are black or brown in color with elongated, slender bodies and short wing covers. Adults vary in size depending on the species. Rove beetles are quick runners and flyers, and many curl the tip of their abdomen slightly upward as they run, or when disturbed by potential predators.

COMMON PREY: Insect eggs and small larvae, slugs, and mites

SPECIES IN NORTH AMERICA: Approximately 3,000

DEVELOPMENT TIME: 20 days to a year

GENERATIONS PER YEAR: One to multiple (varies with species)

EGG-LAYING SITES: On leaves or under leaf litter

OVERWINTERING: Under bark or vegetation as larvae, pupae, or adults

ADDITIONAL FOOD SOURCES: Dead organic matter

ADDITIONAL HABITAT: Eggs are laid on leaves or under debris, and both larvae and adults are found in soils, mulch, leaf litter, or on vegetation.

CONSERVATION STRATEGIES: Rove beetles may benefit from beetle banks (banked areas within fields planted with dense bunch grasses). Researchers have found that rove beetle species diversity and abundance increased with the age of wildflower field borders. Limit tillage and soil fumigants, which alter habitat both for larvae and adults. Rove beetles will use mulch, litter, compost piles, or rock piles for overwintering habitat. Staggering harvest times of perennial forage crops such as alfalfa or clover may encourage these beetles to remain in fields rather than migrate elsewhere.

Flower Flies, Hover Flies

ORDER: Diptera

FAMILY: Syrphidae

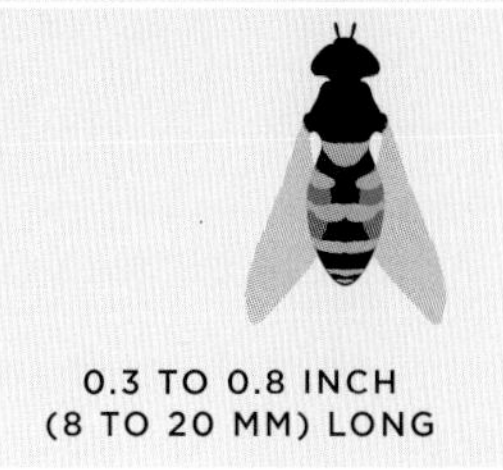

0.3 TO 0.8 INCH
(8 TO 20 MM) LONG

Flower Fly adult

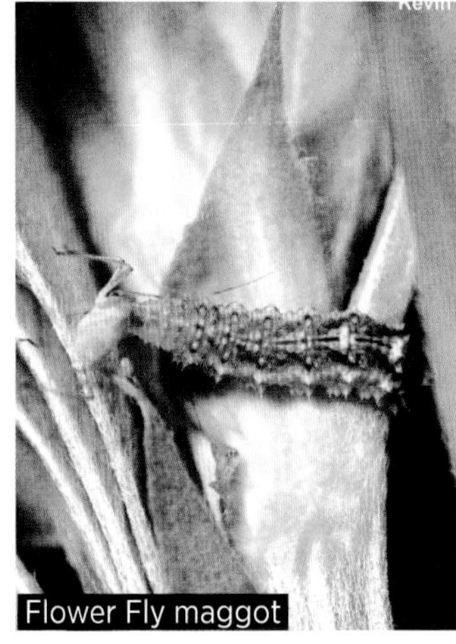
Flower Fly maggot

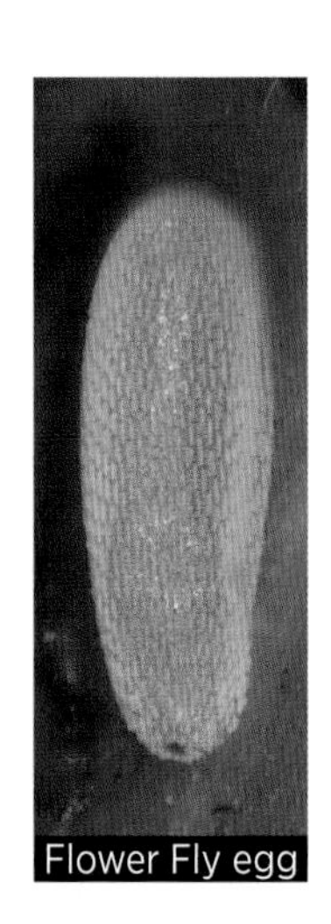
Flower Fly egg

FLOWER FLY EGGS are cream-colored ovals laid on foliage among prey. These develop into brown-gray-green legless larvae, some of which bear distinctive markings, stripes, or spines. Larvae swing their heads from side to side while hunting prey in leaf litter or on foliage. Adult flies have two wings, short, stout antennae, and large, broad eyes. Adults often have bright coloration, and many species mimic the coloration of bees or wasps, some to a striking degree.

COMMON PREY: Flower fly larvae will eat aphids, scales, mealybugs, spider mites, and thrips. They have been known to consume as many as 50 aphids per day, and are particularly important in controlling aphid infestations early in the season, when cooler temperatures may inhibit other predators.

SPECIES IN NORTH AMERICA: Approximately 900

DEVELOPMENT TIME: Four to 10 weeks

GENERATIONS PER YEAR: One to multiple (varies with species)

EGG-LAYING SITES: Eggs are laid singly or in small clumps on foliage near prey (e.g,. next to an aphid colony).

OVERWINTERING: In leaf litter or in the soil as larvae, pupae, or adults

ADDITIONAL FOOD SOURCES: Adults eat pollen and nectar.

ADDITIONAL HABITAT: Adults can be found in habitats with abundant flowering plant.

CONSERVATION STRATEGIES: Protect grasslands, rangelands, meadows, gardens, field borders, hedgerows, and plantings within crop fields. When planted or managed to provide a continuous bloom of flowering plants, these areas can support large flower fly populations. Shallow, open flowers, such as plants in the willow, rose, buttercup, aster, and carrot families, may be especially attractive to flower flies. Since flower flies are less active under windy conditions, windbreaks may provide shelter and increase populations. Limit tilling and burning, which may destroy overwintering sites.

Predatory Wasps

ORDER: Hymenoptera

FAMILY: Vespidae, Sphecidae

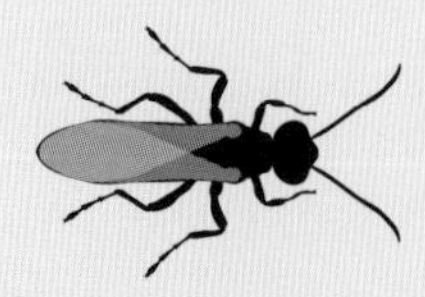

VESPID: 0.4 TO 1 INCH (10 TO 25 MM) LONG;
SPHECID: 0.4 TO 1.25 INCHES (10 TO 30 MM) LONG

Predatory Sphecid Wasp adult

ADULT VESPID WASPS often have a thin waist and are black or brown with white, yellow, red, or orange markings. Vespid wasps fold their wings together when at rest, and at a glance appear to have only one thin pair of wings. Adults of sphecid wasps have a very thin, elongated waist, and are fully black, slightly metallic, or black with red, yellow, or white markings. Sphecids tend to be more slender than vespid wasps.

COMMON PREY: Adult females collect prey to bring back to their nests as food for their larvae. Some species are generalists, feeding on caterpillars, beetles, flies, or true bugs, while others may hunt more selectively on particular pest groups such as grasshoppers or aphids.

SPECIES IN NORTH AMERICA: Approximately 320 of Vespidae, 1,150 of Sphecidae

DEVELOPMENT TIME: Approximately 40 days to a year (varies with species)

GENERATIONS PER YEAR: One to several (varies with species)

EGG-LAYING SITES: Eggs are laid in a chamber within a nest.

OVERWINTERING: Within nest as prepupae or adults

ADDITIONAL FOOD SOURCES: Adult wasps feed primarily on nectar, though some species also feed on rotting fruit or the juices of prey.

ADDITIONAL HABITAT: These wasps build nests in which their young develop. While all sphecid wasps are solitary, with each individual female constructing and provisioning her nest, some vespids are social, forming colonies with a queen and a division of labor among workers. Social vespids construct paper nests made from wood chewed to pulp, and feed masticated prey to their young. Social vespids can be aggressive, particularly the yellowjackets: however, paper wasps (in the genus *Polistes*) are less aggressive and may be worth encouraging because they are excellent caterpillar hunters. Solitary vespids construct cells out of clay or chewed foliage on twigs, stems, crevices of walls, or between rocks. Many solitary sphecid wasps build nests in cavities or in the ground, and may utilize pieces of grass, mud, or resin in construction of their nest.

CONSERVATION STRATEGIES: As adults, predatory wasps drink nectar from shallow flowers such as milkweeds, and members of the sunflower, carrot, and mint families. Some species of solitary wasps will readily use artificial nesting boxes (such as those used to attract solitary bees). Similarly, box shelters with open bottoms erected on posts around tobacco fields have been shown to encourage nesting by paper wasps and reduce pest caterpillars.

Parasitoid Wasps

ORDER: Hymenoptera

SUPER FAMILY: Ichneumonoidea, Chalcidoidea

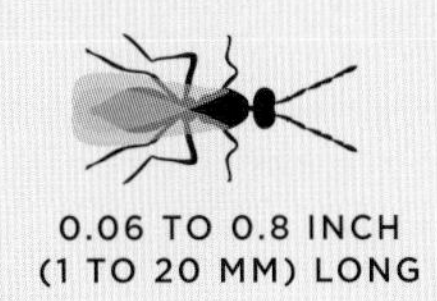

0.06 TO 0.8 INCH (1 TO 20 MM) LONG

AS ADULT WASPS, some species are extremely tiny while others are bigger (up to 1 inch (2.5 cm) in length). All have slender bodies with narrow waists. Females have an ovipositor, a long, stingerlike appendage used to deposit eggs into hosts, though it is less visible in some species. Larger parasitoid wasps in the superfamily Ichneumonoidea (which includes the families Ichneumonidae and Braconidae) are dark with red, orange, or yellow markings, and long, threadlike antennae. Tiny parasitoid wasps in the superfamily Chalcidoidea (which includes families such as Aphelinidae, Trichogrammatidae, Encyrtidae, and Chalcididae) are black, dark blue, or green, and often metallic in color.

COMMON HOSTS: Many parasitoid wasps are host specific and highly effective in regulating the populations of specific pests. Hosts include eggs, nymphs, larvae, or adults of aphids, whiteflies, scales, caterpillars, flies, beetles, leafhoppers, stink bugs, and many other insects. The life cycle of parasitoid wasps is closely synchronized to that of their hosts. An adult female wasp finds a host at the appropriate life stage and deposits one or several eggs on, inside, or near the host. The larvae develop on or inside the host, feeding on it but usually not killing it until they reach maturity and pupate. Adult wasps emerge and seek new hosts to repeat the cycle.

SPECIES IN NORTH AMERICA: Approximately 5,000 in the superfamily Ichneumonidea; approximately 2,600 in the superfamily Chalcidoidea

DEVELOPMENT TIME: Varies with species/host

Parasitoid Wasp adult

GENERATIONS PER YEAR: Numerous, overlapping

EGG-LAYING SITES: Eggs are laid on the surface of host's body or inserted inside host's body.

OVERWINTERING: As an egg or larva within their host, as a pupa within their cocoon, or as adults

ADDITIONAL FOOD SOURCES: Adults are free-living and feed on nectar, as well as extrafloral nectar, honeydew, and occasionally pollen.

ADDITIONAL HABITAT: Adults can be found on flowers with shallow nectar reserves.

CONSERVATION STRATEGIES: Wooded edges and hedgerows near fields, in combination with nectar sources, increase the abundance of parasitoids and levels of parasitism. Permanent plantings with a succession of flowering plants that bloom throughout the season, including species in the carrot, legume, aster, and mint families, will support adult parasitic wasps and increase their longevity and reproduction. Increasing crop diversity can also increase parasitic wasps on farms.

Scarab-Hunting Wasps

ORDER: Hymenoptera

FAMILY: Scoliidae, Tiphiidae

SCOLIID: 0.8 TO 2 INCHES (20 TO 50 MM) LONG; TIPHIID: 0.3 TO 1 INCH (6 TO 25 MM) LONG

Scoliid Wasp adult

Tiphiid Wasp adult

ADULT SCOLIID WASPS are large, robust wasps that have spiny, bristly legs. Their bodies are often black with bright orange, yellow, or red colorations, and their wings are darkened or metallic blue-black. Tiphiid wasps have slender bodies and elongated abdomens. Most are black, though some are black with yellow markings. Females of some tiphiid species may be wingless.

COMMON PREY: These wasp larvae are parasitoids of beetle larvae, particularly scarab beetles, including Japanese beetles. The adult female wasps search fields and lawns for beetle grubs, and when they detect them, dig down through the soil to the grub, paralyze it, lay an egg on it, and leave the grub for their young to devour. Sometimes these wasps will paralyze grubs without leaving eggs, and grubs do not recover from the sting.

SPECIES IN NORTH AMERICA: Approximately 20 species of Scoliidae; approximately 140 species of Tiphiidae

DEVELOPMENT TIME: 10 to 12 months between egg deposition and emergence of adult

GENERATIONS PER YEAR: One

EGG-LAYING SITES: Eggs are laid next to paralyzed prey.

OVERWINTERING: Pupae overwinter in soil.

ADDITIONAL FOOD SOURCES: Adults of both families drink nectar; some scoliids will consume pollen

ADDITIONAL HABITAT: Flowering plants

CONSERVATION STRATEGIES: Field borders or permanent plantings with plenty of summer-to-fall-blooming wildflowers can provide adult wasps with the nectar they need to survive and reproduce. Incorporating nectar-producing flowers into the landscape can increase the rate of parasitism of beetle grubs.

Tachinid Flies

ORDER: Diptera

FAMILY: Tachinidae

Tachnid Fly adults

ADULTS OF TACHINID FLIES vary in size and resemble a housefly in general appearance, but with stiff bristles on the abdomen. Coloration varies widely, although many are gray or brown with dark bristles; others have vivid yellow or red markings or are metallic blue or green. Most species of tachinid flies attack the larval stage of their host, and the developing fly consumes the host and pupates around the time the host dies.

COMMON HOSTS: These parasitoids target the larval stage of certain butterflies, moths, beetles, sawflies, true bugs, and grasshoppers, while others attack a variety of arthropod hosts

SPECIES IN NORTH AMERICA: Approximately 1,000

DEVELOPMENT TIME: Varies with species, but can be less than four weeks

GENERATIONS PER YEAR: Multiple

EGG-LAYING SITES: Eggs are laid either near their host or directly on their host, and a few species insert eggs into the host's body.

OVERWINTERING: As larvae or pupae within host, or as pupae or adults in soil or under leaf litter

ADDITIONAL FOOD SOURCES: Adults are free-living and feed on nectar and pollen.

ADDITIONAL HABITAT: Some species will pupate in leaf litter; others pupate within their host. Adults can be found on flowers.

CONSERVATION STRATEGIES: To support tachinids, maintain consecutively blooming wildflowers for a steady supply of nectar as an adult food source. Flowers with easily accessible nectar are best. Plants from the carrot, aster, rose, willow, mint, and milkweed families may be particularly attractive to these flies. Tachinids are more abundant in less disturbed field margins, and it may require two or more years for their populations to recover following a habitat disturbance.

Jumping Spiders, Wolf Spiders, Orb Weaver Spiders, Sheet-Weaving Spiders

ORDER: Araneae

FAMILY: Lycosidae, Salticidae, Araneidae, Linyphiidae

SPIDERS HAVE FOUR PAIRS of walking legs and two body regions known as the cephalothorax (the head and thorax combined) and abdomen. Silk-spinning organs are found at the posterior end of the abdomen. Wolf spiders have brown, black, or dirty yellow coloration, often with one or more longitudinal stripes on their back. Jumping spiders have variable body coloration, usually dark with iridescent patterns, a fuzzy appearance, and large, forward-facing eyes. Orb-weavers tend to be dark, with bright patterns of yellow, green, or red. Their webs are large vertical spiral webs suspended between vegetation. Sheet-weaving spiders are small, dark brown or black spiders that make semihorizontal webs that form a small sheet between vegetation. Sheet-weaving spiders can disperse long distances and colonize new crop fields by ballooning, letting themselves be carried by wind on their silk threads.

COMMON PREY: Spiders are often the most abundant and diverse predators in agricultural fields. Spider communities can be important in stabilizing populations of insect pests such as beetles, caterpillars, leafhoppers, and aphids. Web-building spiders catch more prey than they can consume. However, spiders are generalist predators and may feed indiscriminately on other insects, including beneficial species.

SPECIES IN NORTH AMERICA: Of the approximately 3,500 species of spiders found in North America, about 240 species are wolf spiders, 315 species are jumping spiders, 160 species are orb weaver spiders, and at least several hundred species are sheet-weaving spiders.

Orb Weaver Spider

DEVELOPMENT TIME: Several weeks, but adults can live for two to three years

GENERATIONS PER YEAR: One

EGG-LAYING SITES: Spider eggs may be laid within silken sacs in leaf litter, attached to the web, or attached to the the body of their mother.

OVERWINTERING: Adults or eggs overwinter in silken nests in the soil, grass clumps, plant debris, under bark, or inside hollow stalks of vegetation.

ADDITIONAL FOOD SOURCES: A few species may consume some nectar in addition to prey.

ADDITIONAL HABITAT: Spiders will live anywhere with abundant prey, including forests, grasslands, urban settings, riparian areas, and in

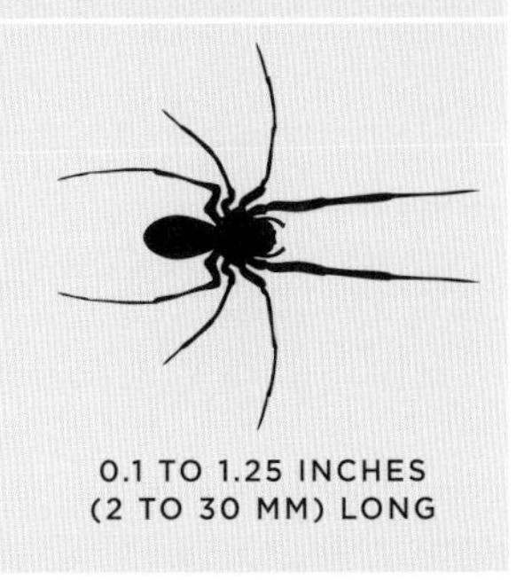

Jumping Spider

Wolf Spider

farm systems, where they live in the crop canopy or on the soil surface. Habitat structure is important for spiders, particularly orb-weaver or sheet-weaving spiders, which construct webs to ensnare prey. Others, such as wolf spiders and jumping spiders, actively pursue prey on the ground or on plant foliage.

CONSERVATION STRATEGIES: While many families of spiders can be found in crop settings, wolf, jumping, sheet-weaving, and orb weaver spiders are some of the most common and important groups. Spider populations can be disrupted by planting and harvesting of crops, but they can recolonize fields if suitable habitats such as natural areas or permanently planted field borders are nearby. Field borders, brush piles, or hedgerows can provide spiders with shelter, as do fields or orchards with ground covers or crop residue. Maintaining cover crops, staggering the harvest of perennial field crops, and leaving residue from barley or rye crops can support spiders. Minimum-tillage practices and leaving stubble between planting seasons will provide shelter and may increase the diversity of ground hunting spiders. In experimental vegetable gardens, the inclusion of mulch alone or both mulch and flowering plants increased spider abundance and decreased damage by pests. Recent evidence suggests that some foliage-wandering spiders, such as jumping spiders, may commonly consume some nectar and that nectar may contribute to their fitness. Flowers that provide floral or extrafloral nectar may be more important to some spiders than previously thought.

Harvestmen

ORDER: Opiliones

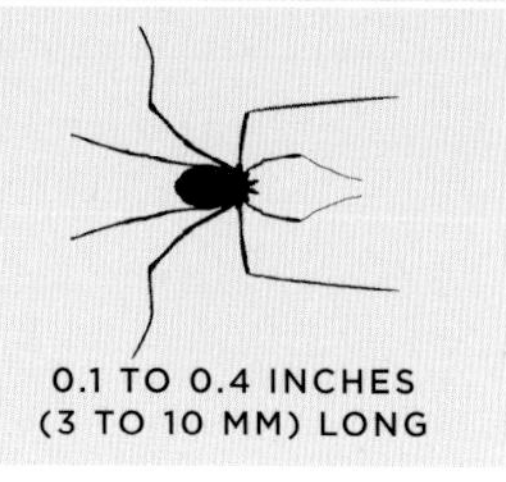

0.1 TO 0.4 INCHES (3 TO 10 MM) LONG

Harvestmen adults

HARVESTMEN have two body regions that are broadly joined and appear as one. Their oval bodies may have gray, brown, or tan color patterns. They have eight exceptionally long, stilt-like legs, and use the second pair to search for prey and sense their environment.

COMMON PREY: Harvestmen feed on a variety of insects, including true bugs, beetles, and eggs of moths.

SPECIES IN NORTH AMERICA: Approximately 150

GENERATIONS PER YEAR: One

EGG-LAYING SITES: Eggs are laid in soil or leaf litter.

OVERWINTERING: Eggs overwinter in leaf litter or soil.

ADDITIONAL FOOD SOURCES: Some are omnivorous, and will feed on dead plant or animal material.

ADDITIONAL HABITAT: Harvestmen can be found in gardens, field edges, grasslands, and woodlands.

CONSERVATION STRATEGIES: Limit tillage, which can harm harvestmen populations. Retain brush or straw piles as shelter.

Predatory Mites

ORDER: Acari

FAMILY: Phytoseiidae

0.06 INCH (2 MM) LONG OR LESS

PREDATORY MITES have pear-shaped bodies and are pale to red in color, though some species take on the color of their prey after they feed. Mite larvae have six legs, while adults have eight.

COMMON PREY: Spider mites and small insects such as thrips, mealybugs, psocids, or whiteflies. The Phytoseiidae family is of particular importance to biocontrol, but other families such as Laelapidae also have members that control small caterpillars and fungus gnats.

SPECIES IN NORTH AMERICA: Approximately 400

DEVELOPMENT TIME: Several days to several weeks (depending on weather and species)

GENERATIONS PER YEAR: 10 or more

EGG-LAYING SITES: On leaves or flowers, near groups of prey

OVERWINTERING: Adults overwinter on trees or in soil debris.

ADDITIONAL FOOD SOURCES: Pollen or fungal spores

CONSERVATION STRATEGIES: Maintain cover crops, ground covers, or permanent plantings that offer alternative prey and shelter. Mites have been observed to move from cover crops into fields infested with their prey. Despite their relatively rapid rate of reproduction, it can take time to build up populations of predatory mites large enough to control pests, and populations can be quickly decimated by pesticide applications. Predatory mites are usually susceptible to the same pesticides to which spider mites and thrips have developed resistance, so avoid nonselective spraying.

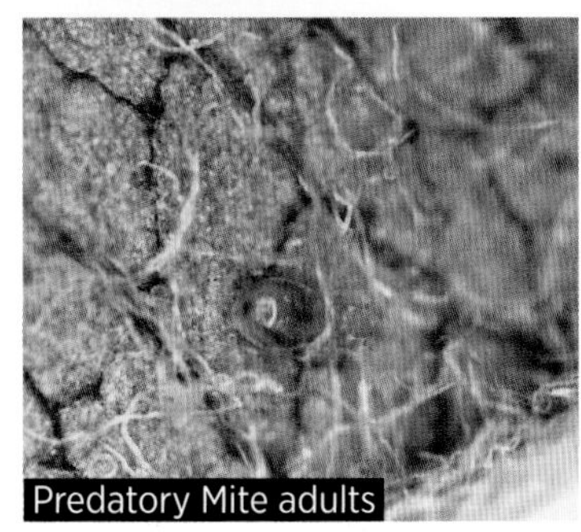
Predatory Mite adults

PART 5

Plants for Conservation Biocontrol

CONSERVATION BIOCONTROL can provide many services to a farm or garden, and it all starts with the plants. Beyond their many functional roles — such as stabilizing soil, reducing runoff of nutrients and water, removing or trapping air pollutants, and providing habitat to support biodiversity — native plants can beautify land. From bunch grasses to trees, plants provided for beneficial insects can transform your farm or garden into a landscape teeming with life. Here we include profiles of perennial wildflowers, shrubs and trees, and plants suitable for cover crops or insectary plantings, to help you find species that are most suited to your habitat needs and land.

*With readily accessible pollen and nectar and large flowers where beneficial insects can perch, sunflowers (*Helianthus *spp.) are one of many groups of wildflowers that can be planted to support beneficial insects.*

THE PLANTS LISTED in this section are some of the species recognized in published research, and in the authors' own observations, as being particularly good for supporting the food and shelter requirements of predatory and parasitoid insects. Many other species not listed here may provide equal or even better resources for various beneficial insects, but currently more research is needed to identify what those plant species are, and how they can be integrated into farm systems.

Wherever possible, we recommend locally native plant species as the best option for supporting locally native beneficial insects. Most of these plants prefer full sun, but vary widely in their moisture requirements. In addition to native species, we provide a short list of low-cost, primarily nonnative plants that can be used for temporary mass insectary plantings or cover crops.

For more information about the range and preferred site characteristics of each of these species, we recommend the USDA Natural Resources Conservation Service PLANTS Database (online at http://plants.usda.gov) as well as the Beneficial Insect Collection Database available through the Lady Bird Johnson Wildflower Center (online at www.wildflower.org/collections).

Native Wildflowers

Because most beneficial insects feed on pollen and nectar as a supplementary food source, native wildflowers are the cornerstone of conservation biocontrol. Just as wild beneficial insects are highly adapted to native wildflowers, the converse is also true: pest insects are typically most attracted to crops or weeds, rather than native wildflowers. Thus, you can be relatively confident that most of the plants featured in this section will not harbor most crop pests. Many more native wildflowers beyond those featured here are undoubtedly extremely valuable for conservation biocontrol. This list simply highlights those that are commercially available and widely recognized for their high attractiveness to beneficials.

Aster (*Symphyotrichum* spp.)

Asters are some of the latest-blooming plants found in many regions of North America, providing an important supply of pollen and nectar long after summer-blooming species have shed their flowers. Many species are rather tall, leggy plants that need equally tall surrounding vegetation to support them. Because of their growth habit, asters are best in native plant field borders, hedgerow edges, riparian buffers, and other places on the farm where they can sprawl onto other vegetation.

LIFE CYCLE: Perennial
RANGE: Nationwide U.S. and Canada
FLOWERING PERIOD: Autumn

Blanketflower/Indian Blanket (*Gaillardia* spp.)

The seed of both annual and perennial blanket-flower species is readily available for most regions of North America. Most species are highly drought tolerant, easy to establish, and attractive to many different beneficial insect species. Perennial species have conservation biocontrol value in native plant field borders, while annual species may be useful in temporary insectary strips.

LIFE CYCLE: Annual/perennial

RANGE: Nationwide U.S. and Canada

FLOWERING PERIOD: Summer

Boneset (*Eupatorium perfoliatum*)

Boneset prefers wet soils and is best suited for use in drainage ditches, farm pond and wetland edges, and other routinely wet sites. Many types of beneficial insects are attracted to the flowers for pollen and nectar.

LIFE CYCLE: Perennial

RANGE: U.S. and Canada east of the Rocky Mountains

FLOWERING PERIOD: Summer

Canada Anemone (*Anemone canadensis*)

Canada anemone is a low-growing wildflower that grows in both sun and shade. Its small flowers attract various beneficial flies and solitary wasp species. It is appropriate for use in native plant field borders and as an understory plant in hedgerows.

LIFE CYCLE: Perennial

RANGE: Midwestern and northeastern U.S.; eastern and central Canada

FLOWERING PERIOD: Spring

Cupplant, Compass Plant, Rosinweed (*Silphium* spp.)

Plants in the *Silphium* genus often reach over 7 feet (2 m) in height, supporting sunflowerlike flowers on thick, pithy stems. Cupplant (*Silphium perfoliatum*) in the Midwest, and starry aster (*Silphium asteriscus*) in the South both work especially well in native plant field borders and herbaceous windbreaks. The hollow stems of cupplant provide nest tunnels for solitary wasps. *Silphium* flowers are attractive to a wide range of parasitoids and generalist insect predators, such as soldier beetles, and often attract and support populations of a red aphid species that does not feed upon crop plants. Those aphids may provide alternative prey for predators and parasitoids when pest aphids are not present.

LIFE CYCLE: Perennial

RANGE: Midwestern, northeastern, and southeastern U.S.

FLOWERING PERIOD: Summer

Daisy Fleabane (*Erigeron* spp.)

Various fleabane daisies are extremely common weedy wildflowers throughout North America. They are most common on poor soils and disturbed sites. Because of their widespread distribution, there is little point in purposefully planting them, and indeed seed for many species is not available. However, because they attract various beneficial insects, especially large numbers of small flower flies, they are worth tolerating in field borders, and other out-of-the way areas where they are not potential field crop weeds.

LIFE CYCLE: Annual/biennial

RANGE: Nationwide U.S. and Canada

FLOWERING PERIOD: Summer

Dotted Mint (*Monarda punctata*)

Dotted mint produces large amounts of nectar and is commonly visited by many species of solitary wasps. It is relatively easy to establish from seed, is suitable for native plant field borders, and is a possible candidate species for consideration in perennial mass insectary plantings. It is relatively drought tolerant and thrives in many types of soils. A related species, lemon beebalm (*Monarda citriodora*) is a native annual of the southern plains, and is highly attractive to many wasp species. Its seed is relatively low cost, making it an excellent plant for temporary insectary mixes.

LIFE CYCLE: Perennial

RANGE: Midwestern and southern U.S.

FLOWERING PERIOD: Summer

Globe Gilia (*Gilia capitata*)

Gilia is a low-cost species suitable for insectary plantings and native plant field borders. It is relatively easy to start from seed, tolerates a range of soil conditions, and is attractive to many different species of nectar-drinking beneficial insects.

LIFE CYCLE: Annual

RANGE: California, Oregon, and the Southwest

FLOWERING PERIOD: Spring/Summer

Golden Alexanders (*Zizia aurea*)

The umbel-shaped flowers of golden Alexanders provide nectar for small insects with short mouthparts such as various flies and parasitoid wasps. Golden Alexanders tolerates a range of soil conditions, preferring periodically wet locations. It grows prolifically in ditches, the edges of grassed waterways, and in native plant field borders.

LIFE CYCLE: Perennial

RANGE: U.S. and Canada east of the Rocky Mountains

FLOWERING PERIOD: Spring

Goldenrod (*Solidago* spp., *Oligoneuron* spp., *Euthamia* spp.)

Many species of goldenrod occur throughout North America, with wide adaptations to different soil types. Most species have a reputation for attracting numerous beneficial insects to their flowers. Depending on the species, various goldenrods will thrive in native plant field borders, along drainage ditches and in buffer systems, and possibly interplanted into beetle banks.

LIFE CYCLE: Perennial

RANGE: Nationwide U.S. and Canada

FLOWERING PERIOD: Late summer to autumn

Indian Hemp/Dogbane

(*Apocynum cannabinum*)

Dogbane adapts to nearly any soil condition and can form large, aggressive colonies, giving it a reputation for being weedy. Despite this, dogbane flowers attract huge numbers of insect visitors including many beneficial flies and solitary wasps. It is best used in native plant field borders and low-maintenance buffer systems where tough, spreading plants are desired.

LIFE CYCLE: Perennial

RANGE: Nationwide U.S. and Canada

FLOWERING PERIOD: Summer

Ironweed (*Vernonia* spp.)

Ironweeds are generally over 6 feet (2 m) tall, with attractive purple flowers that provide an abundance of pollen and nectar. In addition to flower-visiting insects, ironweeds often attract aphids that can provide an alternative food source for various predators and parasitoids when pests are not present in crop fields. Although soil requirements vary, many ironweeds prefer slightly damp soils, and are well adapted to drainage areas and riparian buffers.

LIFE CYCLE: Perennial

RANGE: Midwestern, eastern, and southeastern U.S.

FLOWERING PERIOD: Summer

Lanceleaf Coreopsis

(*Coreopsis lanceolata*)

Lanceleaf coreopsis is a low-growing flower that thrives in dry, sandy soils with low fertility. In the absence of competing plants, it will spread to form low-growing rhizomatous colonies. Many flies, small solitary wasps, and beetle species are attracted to the shallow flowers. The plant functions well in native plant field borders, many types of farm buffers, and may work as a component of beetle banks or grassed waterways that are infrequently mowed.

LIFE CYCLE: Perennial

RANGE: Native to the eastern U.S. from the Great Lakes south to Texas and Florida; naturalized along the West Coast

FLOWERING PERIOD: Late spring to early summer

Lemon Beebalm

(*Monarda citriodora*)

Lemon beebalm produces large amounts of nectar and is commonly visited by many species of solitary wasps. It is somewhat drought tolerant, low cost, and easy to establish from seed, making it suitable for mass insectary plantings.

LIFE CYCLE: Annual

RANGE: Southern U.S. from Florida to Arizona

FLOWERING PERIOD: Summer

Meadowfoam (*Limnanthes* spp.)

Two species of meadowfoam tend to be occasionally available on a commercial basis. Poached egg plant (sometimes called Douglas meadowfoam, *Limnanthes douglasii*) is the most readily available of the two. White meadowfoam (*Limnanthes alba*) is used as a commercial source of plant oil and its seed tends to be closely guarded by the oilseed industry, although wild-sourced seed is occasionally available. Both meadowfoams bloom early in the year and often can reseed themselves, especially at seasonally wet locations such as vernal pools and drainage areas. Meadowfoam attracts an abundance of beneficial insects, especially flower flies.

LIFE CYCLE: Annual

RANGE: California and the Pacific Northwest

FLOWERING PERIOD: Spring

Milkweed (*Asclepias* spp.)

Research conducted at Washington State University has demonstrated that locally native milkweed species are among the most attractive flowers available to various beneficial insects in the state's eastern vineyards. The huge diversity of milkweed species includes plants adapted to a wide range of soil conditions. The tall stature of many milkweeds makes them most appropriate for use in native plant field borders, or in out-of-the-way buffer systems that do not require routine mowing. Note that while several species are toxic to livestock, grazing animals typically avoid them when other forage is available.

LIFE CYCLE: Perennial

RANGE: Nationwide U.S. and Canada

FLOWERING PERIOD: Summer

Mountain Mint (*Pycnanthemum* spp.)

Mountain mints produce an abundance of nectar during flowering from shallow blossoms, and thus attract abundant diverse insects including many solitary wasps, beneficial flies, and predatory beetles. Most species adapt well to a variety of soil conditions, making them ideal plants for inclusion in native plant field borders, and possibly interplanted into grassy beetle banks.

LIFE CYCLE: Perennial

RANGE: Midwest to eastern Canada and south to Florida; one species native to California

FLOWERING PERIOD: Summer

Partridge Pea
(*Chamaecrista fasciculata*)

Partridge pea produces small droplets of nectar at the base of leaf petioles. These small droplets, and the flowers themselves, attract large numbers of small flies, wasps, ants, bees, and velvet ants, a kind of wingless wasp. Partridge pea is an annual species that does not persist in areas dominated by established perennial plants. It does, however, form an extensive short, spreading canopy, similar to that of hairy vetch, on newly disturbed bare ground. Partridge pea can be used in newly established native plant field borders, or in mass insectary plantings, and likely has value as a warm-season cover crop species.

LIFE CYCLE: Annual

RANGE: Midwestern and southern U.S.

FLOWERING PERIOD: Summer

Phacelia (*Phacelia* spp.)

Most species of phacelia provide an abundance of pollen and nectar over a prolonged bloom period. Lacy phacelia (*Phacelia tanacetifolia*) is a widely available, low-cost species that readily reseeds itself in warm climates and performs well in mass insectary plantings, native plant field borders, and as a cover crop. Flower flies and various parasitoid wasps (along with bees) are common visitors to lacy phacelia. Note that in some cases, lacy phacelia has been suspected of attracting lygus bug pests, and increasing their populations. Where this is a concern, avoid planting it near lygus-susceptible crops, such as strawberries.

LIFE CYCLE: Perennial/annual

RANGE: Native to the western U.S.

FLOWERING PERIOD: Spring to early summer

Plains Coreopsis (*Coreopsis tinctoria*)

Plains coreopsis attracts various beneficial flies (e.g., hover flies and tachinids), as well as solitary wasps. Its primary value comes from its drought tolerance and low-cost seed. It will rarely persist among perennial grasses and wildflowers, but the very low cost of seed makes it useful for mass insectary plantings, and to provide a rapid source of pollen and nectar in native plant field borders while slower-growing perennials are getting established.

LIFE CYCLE: Annual

RANGE: Native to the Great Plains, now naturalized across the U.S. and Canada

FLOWERING PERIOD: Summer

Purplestem Angelica
(*Angelica atropurpurea*)

Purplestem angelica strongly prefers wet soils, and is best suited to the edges of drainage ditches, farm ponds, or other wetlands. Its shallow, umbel-shaped flowers attract large numbers of flower flies and small solitary wasps.

LIFE CYCLE: Perennial

RANGE: Midwest, eastern U.S., and Canada

FLOWERING PERIOD: Spring to summer

Rattlesnake Master
(*Eryngium yuccifolium*)

The striking silver flowers of rattlesnake master attract a variety of beneficial flies and small wasps. The tough foliage of the plant and its extensive root system, along with a tolerance for both dry meadows and wetland edges, make it a useful species for drainage ditches and other nonmowed buffer systems, as well as native plant field borders.

LIFE CYCLE: Perennial

RANGE: Midwestern and southeastern U.S.

FLOWERING PERIOD: Summer

Self-heal (*Prunella vulgaris*)

Self-heal is broadly distributed across the Northern Hemisphere, and while it is native to North America, many of the individual plants observed, especially in the eastern U.S., are weeds of European origin. Seed of the native North American subspecies is typically only available in the Pacific Northwest. An advantage of self-heal in conservation biocontrol projects is the plant's ability to compete well against other vegetation, especially weedy grasses that choke out other, more fragile, wildflowers. When seeded at a high enough density, self-heal will form robust colonies of plants that tolerate occasional mowing and vehicle traffic. These characteristics make it a suitable plant for orchard and vineyard understories.

LIFE CYCLE: Perennial

RANGE: Nationwide U.S. and Canada

FLOWERING PERIOD: Early summer

Sunflower (*Helianthus* spp.)

Seed of annual and perennial sunflower species is readily available for most regions of North America. Most species are easy to establish, tolerate a wide range of soil conditions and disturbance, and attract many different beneficial insect species. Some especially tall-growing species may be suited for incorporation into hedge-

rows or may be appropriate for filter strips, and others will do well in native plant field borders. The annual sunflower (*H. annuus*) may be a candidate for low-cost insectary plantings.

LIFE CYCLE: Annual/perennial

RANGE: Nationwide U.S. and Canada

FLOWERING PERIOD: Late summer to autumn

Wild Buckwheat (*Eriogonum* spp.)

Dozens of various wild buckwheat species are found across the West, especially in arid areas. All tend to be tough, clump-forming plants that flower prolifically but are slow to establish. They are suitable for native plant field borders, and some of the larger species may be suitable for use as hedgerow plants. For nonnative buckwheat (*Fagopyrum esculentum*), see page 231.

LIFE CYCLE: Annual/biennial/perennial

RANGE: Western and central U.S. and Canada

FLOWERING PERIOD: Spring through autumn

Wingstem (*Verbesina* spp.)

Although sometimes considered weedy, various species of wingstem (also called crownbeard and frostweed, depending on the species) attract extremely large numbers of beneficial insects, especially nectar-drinking predatory wasps. Most species of *Verbesina* prefer damp soils such as those found in drainage ditches and riparian buffers. Unlike many summer-blooming wildflowers, wingstem tends to tolerate partial shade. The seed of wingstem can be tricky to find from commercial sources, but these species should be given more attention for the diversity of predatory insects (and pollinators, such as bees) that they support. The hollow stems of plants in this genus may also provide nesting habitat for various wasps.

LIFE CYCLE: Perennial

RANGE: Midwestern, eastern, and southeastern U.S.

FLOWERING PERIOD: Summer

Yarrow (*Achillea millefolium*)

The shallow, open flower surfaces of yarrow are especially attractive to flower flies, tachinid flies, and small wasps. Yarrow seed is low cost, easy to establish, and the plant is very drought tolerant, thriving in poor soils, even with other competing plants. Yarrow can be incorporated into most of the habitat enhancement strategies described in this book, including native plant field borders, perennial mass insectary plantings, and buffer systems. The plant may also be a promising species for incorporating into beetle banks, and even into grassed waterways, since it tolerates occasional mowing.

LIFE CYCLE: Perennial

RANGE: Nationwide U.S. and Canada

FLOWERING PERIOD: Summer

Native Flowering Trees and Shrubs

Traditional conservation systems like hedgerows and windbreaks can be upgraded for beneficial insects simply by selecting native species that are recognized sources of high-quality pollen and nectar. The species in this section represent a broad cross-section of great plants from across North America. To maximize the benefits for conservation biocontrol, multiple types of woody plants should be combined in conservation plantings to provide a diverse assemblage of species that flower throughout the growing season.

Basswood (*Tilia americana*)

The large size of mature basswood (or linden) trees (up to 100 feet, or 31 m) makes them inappropriate for typical hedgerow applications. Instead, they are best for incorporating in farm reforestation efforts or as home shade trees. The large, showy flowers of basswood produce huge volumes of nectar, attracting many insect species: bees, wasps, flies, beetles, butterflies, and others. Young trees may not flower for several years, making large transplants the best choice for providing nectar resources quickly.

RANGE: Entire U.S. and Canada east of the Rocky Mountains

FLOWERING PERIOD: Late spring

Buffaloberry (*Shepherdia* spp.)

Buffaloberries are tough, thorny prairie shrubs tolerant of extreme summer and winter temperatures as well as highly saline soils, although they prefer slightly wetter soils, or cooler north-facing slopes. They produce separate male and female flowers, on separate plants, in late spring, attracting a variety of beneficial insects. They are useful for slope stabilization and as windbreak plants in prairie climates where few other shrubs will thrive. Buffaloberry reaches 6 feet (1.8 m) in height.

RANGE: Nationwide Canada and U.S., except southeastern states

FLOWERING PERIOD: Late spring

Buttonbush (*Cephalanthus occidentalis*)

Buttonbush is a useful hedgerow and buffer system plant for eastern North America. It is one of the few shrub species in its region to flower in the summer, and is shade tolerant. The showy flowers attract a wide range of beneficial insects. The plant's preference for wet soils makes it useful for stabilizing drainage areas and stream banks, and revegetating wetlands. The typical mature height is 6 feet (1.8 m).

RANGE: Midwestern and eastern U.S. and Canada, south to Texas and Florida

FLOWERING PERIOD: Summer

Coffeeberry (*Rhamnus californica*)

Coffeeberry is a medium-sized (5 feet, or 1.5 m, at maturity) California shrub with good drought tolerance and prolific flowers. It is commonly used in California hedgerows specifically to attract beneficial insects.

RANGE: California

FLOWERING PERIOD: Spring

Coyotebrush, Baccharis (*Baccharis* spp.)

Coyotebrush (*Baccharis pilularis*) has separate male and female shrubs, and only the male plants bear pollen, making them more attractive to some insects. Both male and female plants flower prolifically in late summer, and in some cases through the fall and early winter within their native range, at a time when few other plants are in bloom. This makes them especially valuable to beneficial insects. They are very drought tolerant and well adapted for use in hedgerows. Several other closely related *Baccharis* species are found in other parts of the country; all are also potentially good conservation biocontrol plants. Depending on the species, these shrubs may approach 10 feet (3 m) in height at maturity.

RANGE: Western and southern U.S.

FLOWERING PERIOD: Late summer, autumn

Desert Sweet
(*Chamaebatiaria millefolium*)

Desert sweet grows well under dry, nutrient-poor conditions. It can be found in dry, rocky habitats, and is a useful plant for sites with harsh conditions. This shrub can reach 6 feet (1.8 m), and its profuse white flowers support numerous predatory wasps at a time when many other shrubs have finished flowering.

RANGE: Southwest to western U.S.

FLOWERING PERIOD: Midsummer, early autumn

Elderberry (*Sambucus* spp.)

There are two species of elderberry distributed across most of North America: the sometimes treelike black species (*Sambucus nigra*) and the generally smaller, more shrubby, red species (*Sambucus racemosa*). The abundant, shallow, showy flowers of both species attract small, beneficial insects. The pithy stems of each plant often hollow out during branch dieback, creating nesting cavities for various solitary wasps. Both elderberries tolerate coppice pruning, which involves cutting the entire plant back to the ground to produce new shoots and a shrubby form. It is commonly used in hedgerows in California. Elderberry may be attractive to various leafhoppers and sharpshooters, pest insects of grapes. If this is a concern, avoid planting elderberries adjacent to vineyards. Conversely, this could make elderberry potentially useful as a "banker plant," a noncrop plant that can be managed to provide ongoing pest populations to sustain beneficial parasitoids (see chapter 4). Mature elderberries can reach 20 ft (6 m) in height.

RANGE: Nationwide U.S. and Canada

FLOWERING PERIOD: Late spring, early summer

False Indigo (*Amorpha fruticosa*)

Commonly found along rivers, ponds, or in wet soils, false indigo is well adapted to farm wetland or drainage basin settings. With supplemental irrigation during establishment, it may be appropriate for upland hedgerow sites. False indigo rarely grows above 10 feet (3 m) in height, but produces large purple flowers in great profusion. In some areas it is considered weedy, and should not be further introduced beyond its documented native range. Leadplant (*Amorpha canescens*) is a low-growing related species native to the Midwest and Great Plains.

RANGE: Nationwide U.S. and eastern Canada; introduced extensively beyond its native range.

FLOWERING PERIOD: Late spring

Hawthorn (*Crataegus* spp.)

Hawthorns are common hedgerow trees in Britain, where the dense thorny branches are valued for fencing, and the flowers and berries are recognized for their wildlife value. Several dozen species of hawthorn are native to various parts of North America, and most should respond to periodic coppice pruning by producing clumps of new shrubby trunks. All are likely good candidate species for hedgerows. Native North American hawthorn species should not be confused with the nonnative oneseed hawthorn (*Crataegus monogyna*), which is considered invasive in many areas. Hawthorn is susceptible to the disease fire blight and may be an alternate host plant for the apple maggot. Consider these factors where appropriate.

RANGE: Nationwide U.S. and Canada

FLOWERING PERIOD: Spring

Ocean Spray (*Holodiscus discolor*)

Ocean spray is tolerant of a wide variety of soil conditions (from periodically wet, to droughty, to rocky), and both sun and shade. It is commonly used in beneficial insect hedgerows in California and the Pacific Northwest because of its large, nectar-rich flower clusters that bloom in the summer, when few other nectar sources are available. Ocean spray is susceptible to fire blight.

RANGE: Western U.S. and Canada

FLOWERING PERIOD: Summer

Spirea (*Spiraea* spp.)

Spirea species tend to prefer fertile, slightly wet soils. The long-lasting blooms attract a variety of beneficial insects, especially small flies and wasps. Spireas are useful in hedgerows, bordering grassed waterways, or in wet drainage areas. Since spirea is an alternate host plant for fire blight, it should be used with caution around apple and pear orchards.
RANGE: Nationwide Canada and U.S. except southwestern states and Florida
FLOWERING PERIOD: Spring and summer

Toyon (*Heteromeles arbutifolia*)

Although native to California, toyon has been widely cultivated beyond its native range. It is both drought tolerant and able to survive limited below-freezing temperatures. Toyon is a common species used in native plant hedgerows across California. Because toyon is susceptible to the apple and pear disease fire blight, use it with caution if it is adjacent to commercial orchards.
RANGE: California
FLOWERING PERIOD: Early summer

Wild Lilac (*Ceanothus* spp.)

Although slow to mature, wild lilacs once established produce huge profusions of flowers that attract large numbers of beneficial insect species. Many western wild lilacs are drought tolerant and are widely incorporated into beneficial insect hedgerows across California.
RANGE: California, a few species nationwide U.S. and southern Canada
FLOWERING PERIOD: Spring

Wild Plum and Wild Cherry
(*Prunus* spp.)

Hedgerows and windbreaks are common uses for various wild plum and cherry species, which rarely reach more than 30 feet (9 m) in height. These plants are good potential candidates for coppice pruning, which involves periodically cutting the main trunk back to the ground to encourage suckering and a denser, shrublike form. Most plants in this genus are susceptible to several of the same insect pests as commercial *Prunus* species, making them possible "banker plant" candidates (see chapter 4). Because they are also susceptible to the same diseases as commercial stone fruit, however, incorporate wild plums into orchards with careful planning.

RANGE: Nationwide U.S. and Canada

FLOWERING PERIOD: Spring

Wild Rose (*Rosa* spp.)

Native wild roses are sometimes incorporated into hedgerows or wet drainage areas. Most species are not drought tolerant and prefer wetter soils. Flower flies and various other beneficial insects are attracted to the flowers. However, in areas with significant Japanese beetle populations, rose blossoms may be subject to feeding damage. Wild rose is also susceptible to leafroller caterpillars of several species, making it potentially useful as a **banker plant**, a noncrop plant

that can be managed to provide ongoing pest populations to sustain beneficial parasitoids (see chapter 4).

RANGE: Nationwide U.S. and Canada

FLOWERING PERIOD: Early summer

Willow (*Salix* spp.)

Small nectar-seeking flower flies, parasitoid wasps, and ladybird beetles are common visitors to various native willows, which typically are among the earliest-blooming plant species in any given region. Most willows strongly prefer wet soils and can be planted as live stake cuttings into riverbanks, along ponds, and in wet drainage areas. Most willows also respond well to coppice pruning to develop dense multistem thickets. Willows may be attractive to various leafhoppers and sharpshooters, pest insects of grapes. If this is a concern, avoid planting willows adjacent to vineyards.

RANGE: Nationwide U.S. and Canada

FLOWERING PERIOD: Spring

Native Grasses

Native grasses have conservation value for beneficial insects as shelter for overwintering, pupation, and egg-laying. In fact, the dense crowns of species such as deergrass are routinely identified as important overwintering sites for lady beetles in some parts of the country, with dozens or even hundreds of individual lady beetles observed clustering into a single grass clump on some western farms. Similarly, a common practical use of grasses for conservation biocontrol is in beetle banks to provide winter shelter for predatory ground beetles.

These grasses are also valuable for seeding into field borders with diverse wildflower species. Such mixed grass-wildflower meadows create a more complex habitat and form a tight living root mass that can better resist weed encroachment. Native grasses can also be fully integrated into many types of filter strip systems, and may be appropriate as orchard or vineyard understory plants. Because some native grasses have an aggressive growth habit, be careful when incorporating them into areas where wildflower abundance and diversity are priorities. If this is the case, consider a wildflower-grass seeding ratio of 3:1 or 4:1.

Here is a selection of good grass species for conservation biocontrol efforts.

Big Bluestem (*Andropogon gerardii*)

RANGE: Central and eastern U.S. and Canada

HEIGHT: 7 feet (2.4 m)

WATER REQUIREMENTS: Moderate

Bluebunch Wheatgrass (*Pseudoroegneria spicata*)

RANGE: Western U.S. and Canada

HEIGHT: 4 feet (1.2 m)

WATER REQUIREMENTS: Low

Blue Wild Rye (*Elymus glaucus*)

RANGE: Western U.S. and Canada

HEIGHT: 3.5 feet (1 m)

WATER REQUIREMENTS: Low

California Oatgrass (*Danthonia californica*)

RANGE: Western U.S. and Canada

HEIGHT: 2 feet (0.6 m)

WATER REQUIREMENTS: Moderate

Canada Wild Rye (*Elymus canadensis*)

RANGE: Nationwide Canada and U.S. except southeastern states

HEIGHT: 4 feet (1.2 m)

WATER REQUIREMENTS: Moderate

Deergrass (*Muhlenbergia rigens*)

RANGE: California and southwestern U.S.

HEIGHT: 6 feet (1.5 m)

WATER REQUIREMENTS: Moderate

Big Bluestem

Bluebunch Wheatgrass

Blue Wild Rye

California Oatgrass

Canada Wild Rye

Deergrass

Eastern Gamagrass
(Tripsacum dactyloides)
RANGE: Eastern and southeastern U.S.
HEIGHT: 8 feet (2.4 m)
WATER REQUIREMENTS: High

Idaho Fescue *(Festuca idahoensis)*
RANGE: Western U.S. and Canada
HEIGHT: 2 feet (0.6 m)
WATER REQUIREMENTS: Low

Indian Grass *(Sorghastrum nutans)*
RANGE: Central and eastern U.S. and Canada
HEIGHT: 6 feet (1.8 m)
WATER REQUIREMENTS: Moderate

Little Bluestem
(Schizachyrium scoparium)
RANGE: Eastern and central U.S. and Canada
HEIGHT: 3 feet (1 m)
WATER REQUIREMENTS: Low

Prairie Dropseed
(Sporobolus heterolepis)
RANGE: Central U.S. and Canada
HEIGHT: 3 feet (1 m)
WATER REQUIREMENTS: Low

Prairie Junegrass
(Koeleria macrantha)
RANGE: Nationwide Canada and U.S. except southeastern states
HEIGHT: 2 feet (0.6 m)
WATER REQUIREMENTS: Moderate

Roemer's Fescue *(Festuca roemeri)*
RANGE: West Coast U.S. and Canada
HEIGHT: 2 feet (0.6 m)
WATER REQUIREMENTS: Low

Sideoats Grama
(Bouteloua curtipendula)
RANGE: Central U.S. and Canada
HEIGHT: 4 feet (1.2 m)
WATER REQUIREMENTS: Low

Slender Wheatgrass
(Elymus trachycaulus)
RANGE: Nationwide U.S. and Canada except southeastern states
HEIGHT: 3 feet (1 m)
WATER REQUIREMENTS: Moderate

Wiregrass *(Aristida stricta* and *Aristida beyrichiana)*
RANGE: Southeastern U.S.
HEIGHT: 4 feet (1.2 m)
WATER REQUIREMENTS: Moderate

Eastern Gamagrass

Idaho Fescue
Indian Grass
Little Bluestem
Prairie Dropseed
Roemer's Fescue
Prairie Junegrass
Sideoats Grama
Slender Wheatgrass
Wiregrass

Cover Crops and Nonnative Insectary Plants

The following plants are low-cost flowers that are known to attract large numbers of beneficial insects. Several of these plants, including phacelia, buckwheat, oilseed radish, and various vetches, clovers, and mustards, are all routinely used as summer or winter cover crops. When allowed to flower, they provide a mass source of pollen and nectar, and can support large beneficial insect populations. These and the other species on this list are also suitable for mass insectary plantings as well as insectary rows within crop fields. For diverse insectary plantings with longer bloom times, different species from this list can be seeded together. This approach will provide more variety of flower shapes and sizes and, in time, may support greater beneficial insect species diversity.

Most of the annual species listed below should be seeded in the spring, after the average date of last frost. In warmer climates, fall seeding may also be appropriate.

Alyssum (*Lobularia maritima*)

Alyssum is a low cost, low-growing, and relatively maintenance-free annual flower that is sometimes mass-planted between rows of vegetables to attract various beneficial insects. It should be noted, however, that as a member of the brassica family, this species may attract brassica-feeding pests, including flea beetles and harlequin bugs.

LIFE CYCLE: Annual

FLOWERING PERIOD: Summer

Bachelor Button (*Centaurea cyanus*)

The very low cost of bachelor button makes it suitable for mass planting on field edges to create temporary beneficial insect habitat. Flowers will self-sow under optimal conditions.

LIFE CYCLE: Annual

FLOWERING PERIOD: Summer

Buckwheat (*Fagopyrum esculentum*)

A traditional pseudo-cereal crop, buckwheat develops quickly in warm weather, making it suitable as a summer cover crop. It flowers prolifically, attracting large numbers of beneficial insects. For native buckwheat (*Eriogonum* spp.), see page 218.

LIFE CYCLE: Annual

FLOWERING PERIOD: Summer

Cilantro (*Coriandrum sativum*)

The small flowers of cilantro are highly attractive to many small parasitoid wasps and other beneficial insects. Seed is low cost and widely available.

LIFE CYCLE: Annual

FLOWERING PERIOD: Summer

Clover (*Trifolium* spp.)

Clovers include perennial species such as red, white, and alsike clovers, and annual species such as crimson clover and berseem clover.

LIFE CYCLE: Annual, perennial

FLOWERING PERIOD: Spring to autumn

Dill (*Anethum graveolens*)

Dill's tiny flowers attract many small parasitoid wasps and other beneficial insects. Seed is relatively low cost and widely available.

LIFE CYCLE: Annual

FLOWERING PERIOD: Summer

Forage or Oilseed Radish
(*Raphanus sativus*)

In order to encourage flowering, forage radish should be planted in the spring. Radishes left in the ground are sometimes used to break up soil compaction and as a biofumigant for nematodes. Like mustards, radishes may host pests of brassica crops.

LIFE CYCLE: Biennial

FLOWERING PERIOD: Summer

Hairy Vetch (*Vicia villosa*)

Hairy vetch is one of the most cold-tolerant vetch species and consequently is usually sown in late summer or early fall, even in northern climates, to provide a cool-season cover crop. Hairy vetch is often interseeded with rye to support its trailing and climbing growth habit.

LIFE CYCLE: Annual, biennial, perennial

FLOWERING PERIOD: Spring to autumn

Lacy Phacelia (*Phacelia tanacetifolia*)

This native California wildflower is used extensively as an annual cover crop for both weed suppression and to capture excess soil nitrogen. It produces masses of flowers on a single plant. Note that in some cases, lacy phacelia has been suspected of attracting and increasing the populations of lygus bug pests. Where this is a concern, avoid planting it near lygus-susceptible crops, such as strawberries.

LIFE CYCLE: Annual

FLOWERING PERIOD: Spring to early summer

Mustard (*Brassica* spp.)

Common cover crop species include Indian mustard (*Brassica juncea*) and white mustard (*Sinapis alba*) as well as various types of turnips. Mustards may help suppress weeds and nematodes. Note that some mustard cover crop varieties also support pests and diseases of commercial brassica crops. Consider avoiding their use if this is a concern.

LIFE CYCLE: Annual

FLOWERING PERIOD: Summer

Sunflower (*Helianthus* spp.)

Seed of annual and perennial sunflower species is readily available for most regions of North America. Most species are easy to establish, tolerate a wide range of soil conditions and disturbance, and attract many different beneficial insect species. Some especially tall-growing species may be suited for incorporation into hedgerows or may be appropriate for filter strips, and others will do well in native plant field borders. The annual sunflower (*H. annuus*) may be a candidate for inclusion in low-cost insectary plantings.

LIFE CYCLE: Annual/perennial/biennial

FLOWERING PERIOD: Late summer to autumn

Sweetclover
(*Melilotus officinalis* and *M. albus*)

Most sweetclover varieties do not bloom until the second year after planting. 'Hubam' is a white sweetclover cultivar with an annual life cycle, producing flowers in a single season. It is best adapted to warm, dry climates. Note that sweet clover is often weedy and considered invasive in some areas.

LIFE CYCLE: Biennial

FLOWERING PERIOD: Early summer

PART 6

Appendix

A native grass beetle bank at Grinnell Heritage Farm, in Iowa. This and other habitat features are a cornerstone of this innovative farm's pest management program.

Publications and Fact Sheets

The Xerces Society

www.xerces.org/conservationbiocontrol

Fact sheets and other information. See next page for full list.

Conservation Biocontrol Resources

"Beneficial Insect Habitat in an Apple Orchard — Effects on Pests." Research Brief 71. University of Wisconsin Center for Integrated Agricultural Systems, September 2004. www.cias.wisc.edu

Bugg, Robert L. "Habitat Manipulation to Enhance the Effectiveness of Aphidophagous Hover Flies (Diptera: Syrphidae)." Sustainable Agriculture Research and Education Program, University of California, Winter 1992.

Dufour, Rex. "Farmscaping to Enhance Biological Control." NCAT, 2000. http://attra.ncat.org/attra-pub/summaries/farmscaping.html

Grasswitz, Tess R., and David R. Dreesen. "Pocket Guide to the Beneficial Insects of New Mexico." New Mexico State University Cooperative Extension Service. http://aces.nmsu.edu/ipm/documents/beneficial-insects-booklet-final.pdf

Pesticide Risk Reduction

Adamson, Nancy, Thomas Ward, and Mace Vaughan. "Designed with Pollinators in Mind." *Inside Agroforestry* 20, no. 1 (2012):8, 10. http://nac.unl.edu/documents/insideagroforestry/vol20issue1.pdf

Bentrup, Gary. *Conservation Buffers: Design Guidelines for Buffers, Corridors, and Greenways.* USDA Forest Service, 2008. www.bufferguidelines.net

Hopwood, Jennifer, Scott Hoffman Black, Mace Vaughan, Eric Lee Mader. *Beyond the Birds and the Bees: Effects of Neonicotinoid Insecticides on Agriculturally Important Beneficial Insects.* The Xerces Society for Invertebrate Conservation, 2013. http://www.xerces.org/pesticides

Hooven, L., R. Sagili, E. Johansen. "How to Reduce Bee Poisoning from Pesticides." A Pacific Northwest Extension Publication (PNW 591), Oregon State University, rev. 2013.

"A Whole-Farm Approach to Managing Pests." Sustainable Agriculture Network, 2003. www.sare.org/Learning-Center/Bulletins/A-Whole-Farm-Approach-to-Managing-Pests

Websites

Ohio State University — Agricultural Landscape Ecology Lab

The Agricultural Landscape Ecology (ALE) Lab studies the ecology and management of urban and agricultural habitats, including conservation biocontrol. Numerous free informational resources are available, including insect identification guides.

www.oardc.osu.edu/ale

University of California — Integrated Pest Management Program

An extensive IPM portal with fact sheets on biocontrol, crop-specific pest management guidelines, and pesticide information.

www.ipm.ucdavis.edu

North Carolina State University — Biological Control Information Center

Guidance for insect identification, understanding beneficial insect ecology, and example habitats for conservation biocontrol.

www.ncsu.edu/~dorr/

Farmscaping for Beneficials

Integrated Plant Protection Center, Oregon State University

www.ipmnet.org/BeetleBank/Farmscaping_for_Beneficials.html

Use of cover crops and green manures to attract beneficial insects.

University of Connecticut

http://ipm.uconn.edu/documents/raw2/Use of Cover Crops and Green Manures to Attract Beneficial Insects/Use of Cover Crops and Green Manures to Attract Beneficial Insects.php?display=print

Vineyard Beauty with Benefits

www.wavineyardbeautywithbenefits.com

Developing Vineyard Refugia for Beneficial Insects and Sustainable Pest Management in Washington State, USA.

Native Plants and Ecosystem Services

Michigan State University

www.nativeplants.msu.edu

Books

Altieri, Miguel A., Clara I. Nicholls, and Marlene Fritz. *Manage Insects on Your Farm: A Guide to Ecological Strategies.* Sustainable Agriculture Network, 2005.

Barbosa, Pedro, ed. *Conservation Biological Control.* Academic Press, 1998.

Bellows, Thomas S., and T. W. Fisher, eds. *Handbook of Biological Control: Principles and Applications of Biological Control.* Academic Press, 1999. See especially pp. 319–353.

Flint, Mary Louise, and Steve H. Driestadt. *Natural Enemies Handbook: The Illustrated Guide to Biological Pest Control.* University California Press, 1997.

Gurr, Geoff M., Steve D. Wratten, and Miguel A. Altieri, eds. *Ecological Engineering for Pest Management: Advances in Habitat Manipulation for Arthropods.* CSIRO Publishing, 2004.

Phatak, S. "Managing Pests with Cover Crops." In: *Managing Cover Crops Profitably,* 3rd ed., edited by Andy Clark, 25–33. SAN Handbook Series, book 3. Sustainable Agriculture Network, 2007.

Pickett, Charles H., and Robert L. Bugg. *Enhancing Biological Control: Habitat Management to Promote Natural Enemies of Agricultural Pests.* University of California Press, 1998.

Xerces Society Resources

The Xerces Society for Invertebrate Conservation

855-232-6639

www.xerces.org

Visit our website to order our book (*Attracting Native Pollinators: Protecting North America's Bees and Butterflies.* Storey Publishing, 2011) and to find the following resources:

The Conservation Biocontrol Resource Center, Guidelines for Nest Block Maintenance, Farming for Bees Guidelines, our Pesticide Program, and the Pollinator Conservation Resource Center (an online database of habitat installation guides for beneficial insects, links to information about USDA-administered conservation programs, and more)

References

Bianchi, Felix J. J. A., and Felix L. Wäckers. "Effects of Flower Attractiveness and Nectar Availability in Field Margins on Biological Control by Parasitoids." *Biological Control* 46, no. 3 (2008): 400-408.

Boettner, George H., Joseph S. Elkinton, and Cynthia J. Boettner. 2000. "Effects of a Biological Control Introduction on Three Nontarget Native Species of Saturniid Moths." *Conservation Biology* 14, no. 6 (200): 1798–1806.

References (continued)

Colley, M. R., and J. M. Luna. "Relative Attractiveness of Potential Beneficial Insectary Plants to Aphidophagous Hoverflies (Diptera: Syrphidae)." *Environmental Entomology* 29, no. 5 (2000): 1054–1059.

Collins, K. L., N. D. Boatman, A. Wilcox, J. M. Holland, and K. Chaney. "Influence of Beetle Banks on Cereal Aphid Predation in Winter Wheat." *Agriculture, Ecosystems & Environment* 93, no. 1–3 (December 2002): 337–350.

Cullen, Ross, Keith D. Warner, Mattias Jonsson, and Steve D. Wratten. "Economics and Adoption of Conservation Biological Control." *Biological Control* 45, no. 2 (May 2008): 272–280.

Ellis, J. A., A. D. Walter, J. F. Tooker, M. D. Ginzel, P. F. Reagel, E. S. Lacey, A. B. Bennet, E. M. Grossman, and L. M. Hanks. "Conservation Biocontrol in Urban Landscapes: Manipulating Parasitoids of Bagworm (Lepidoptera: Psychidae) with Flowering Forbs." *Biological Control* 34 (2005): 99-107.

Ferguson, Holly, Sally O'Neal, and Douglas Walsh. "Survey SAYS: Great Grapes!: An IPM Success Story." Extension Bulletin 2025E. Washington State University Extension, 2007.

Fiedler, Anna K., Doug A. Landis, and Steve D. Wratten. "Maximizing Ecosystem Services from Conservation Biological Control: The Role of Habitat Management." *Biological Control* 45 (2008): 254–271.

Frank, Steven D. "Biological Control of Arthropod Pests Using Banker Plant Systems: Past Progress and Future Directions." *Biological Control* 52, no. 1 (2010): 8–16.

Gilliom, Robert J., Jack E. Barbash, Charles G. Crawford, Pixie A. Hamilton, Jeffrey D. Martin, Naomi Nakagaki, Lisa H. Nowell, Jonathan C. Scott, Paul E. Stackelberg, Gail P. Thelin, and David M. Wolock. 2006. "The Quality of Our Nation's Waters: Pesticides in the Nation's Streams and Ground Water, 1992–2001." U.S. Geological Survey Circular 1291, (revised edition, 2007). Available at: http://pubs.usgs.gov/circ/2005/1291/pdf/circ1291_front.pdf.

Gross, K., and J. A. Rosenheim. "Quantifying Secondary Pest Outbreaks in Cotton and Their Monetary Cost with Causal-inference Statistics." *Ecological Applications* 21, no. 7 (2011): 2770–2780.

Harmon, Jason P., Erin Stephens, and John Losey. "The Decline of Native Coccinellids (Coleoptera: Coccinellidae) in the United States and Canada." *Journal of Insect Conservation* 11 (2007): 85–94.

Harwood, J. D., K. D. Sunderland, and W. O. Symondson. "Prey Selection by Linyphiid Spiders: Molecular Tracking of the Effects of Alternative Prey on Rates of Aphid Consumption in the Field. *Molecular Ecology* 13, no. 11 (2004): 3549–3560.

Henneman, M. L., and J. Memmott. "Infiltration of a Hawaiian Community by Introduced Biological Control Agents." *Science* 293, no. 5533 (2001): 1314-1316.

Hodek, I., and A. Honěk. *Ecology of Coccinellidae*. Kluwer Academic Publishers, 1996.

Hoffmann, Michael P., and Anne C. Frodsham. *Natural Enemies of Vegetable Insect Pests*. Cornell Cooperative Extension, 1993.

Isaacs, Rufus, Julianna Tuell, Anna Fiedler, Mary Gardiner, and Doug Landis. "Maximizing Arthropod-mediated Ecosystem Services in Agricultural Landscapes: The Role of Native Plants." *Frontiers in Ecology and the Environment* 7, no. 4 (2009): 196–203.

James, D. G., T. S. Price, L. C. Wright, and J. Perez. "Abundance and Phenology of Mites, Leafhoppers and Thrips on Pesticide-treated and Untreated Wine Grapes in Southcentral Washington." *Journal of Agricultural and Urban Entomology* 19, no. 1 (2002): 45–54.

Jonsson, Mattias, Steve D. Wratten, Doug A. Landis, and Geoff M. Gurr. 2008. "Recent Advances in Conservation Biological Control of Arthropods by Arthropods." *Biological Control* 45 (2008): 172–175.

Koch, R. L. "The Multicolored Asian Lady Beetle, *Harmonia axyridis*: A Review of Its Biology, Uses in Biological Control, and Non-target Impacts." *Journal of Insect Science* 3 (2003): 32.

Landis, Douglas A., Mary M. Gardiner, Wopke van der Werf, and Scott M. Swinton. "Increasing Corn for Biofuel Production Reduces Biocontrol Services in Agricultural Landscapes." *Proceedings of the National Academy of Sciences of the United States of America* 105, no. 51 (2008): 20552–20557.

Landis, Douglas A., Stephen D. Wratten, and Geoff M. Gurr. "Habitat Management to Conserve Natural Enemies of Arthropod Pests in Agriculture." *Annual Review of Entomology* 45 (2000): 175–201.

Lattin, J. D. "Bionomics of the Anthocoridae." *Annual Review of Entomology* 44 (1999): 207–231.

———. "Bionomics of the Nabidae." *Annual Review of Entomology* 34 (1989): 383–400.

Letourneau, Deborah K., and Sara G. Bothwell. "Comparison of Organic and Conventional Farms: Challenging Ecologists to Make Biodiversity Functional." *Frontiers in Ecology and the Environment* 6, no. 8 (2008): 430–438.

Losey, John E., and Mace Vaughan. "The Economic Value of Ecological Services Provided by Insects." *Bioscience* 56, no. 4 (2006): 311–323.

Louda S. M., R. W. Pemberton, M. T. Johnson, and P. A. Follett. "Nontarget Effects: The Achilles' Heel of Biological Control? Retrospective Analyses to Reduce Risk Associated with Biocontrol Introductions." *Annual Review of Entomology* 48 (2003): 365–396.

Lovei, G. L., and K. D. Sunderland. "Ecology and Behavior of Ground Beetles (Coleoptera: Carabidae)." *Annual Reviews Entomology* 41 (1996): 231–256.

Lundgren, Jonathan G. *Relationships of Natural Enemies and Non-prey Foods*. Springer, 2009.

MacLeod, A., S. D. Wratten, N. W. Sotherton, and M. B. Thomas. "'Beetle Banks' as Refuges for Beneficial Arthropods in Farmland: Long-term Changes in Predator Communities and Habitat." *Agricultural and Forest Entomology* 6, no. 2 (2004): 147–154.

McMurtry, J. A., and B. A. Croft. "Life-styles of Phytoseiid Mites and Their Roles in Biological Control." *Annual Review of Entomology* 42 (1997): 291–321

Morandin, L., R. Long, and C. Kremen. "Hedgerows enhance beneficial insects on adjacent tomato fields in an intensive agricultural landscape." Agriculture, Ecosystems & Environment (2014).

Oerke, E. C. "Crop Losses to Pests." *Journal of Agricultural Science* 144, no. 1 (2006): 31–43.

O'Neill, Kevin M. *Solitary Wasps: Behavior and Natural History*. Cornell University Press, 2001.

Prasad, R. P., and William E. Snyder. "Polyphagy Complicates Conservation Biological Control that Targets Generalist Predators." *Journal of Applied Ecology* 43, no. 2 (2006): 343–352.

Prischmann, Deirdre A., and David G. James. "Phytoseiidae (Acari) on Unsprayed Vegetation in Southcentral Washington: Implications for Biological Control of Spider Mites on Wine Grapes." *International Journal of Acarology* 29, no. 3 (2003): 279–287.

References (continued)

Prischmann, Deirdre A., David G. James, and William E. Snyder. "Impact of Management Intensity on Mites (Acari: Tetranychidae, Phytoseiidae) in Southcentral Washington Wine Grapes." *International Journal of Acarology* 31, no. 3 (2005): 277–288.

Prischmann, Deirdre A., David G. James, C. P. Storm, L. C. Wright, and William E. Snyder. "Identity, Abundance, and Phenology of *Anagrus* spp. (Hymenoptera: Mymaridae) and Leafhoppers (Homoptera: Cicadellidae) Associated with Grape, Blackberry and Wild Rose in Washington State." *Annals of the Entomological Society of America* 100, no. 1 (2007): 42–52.

Rabb, R. L., and F. R. Lawson. "Some Factors Influencing the Predation of *Polistes* Wasps on the Tobacco Hornworm." *Journal of Economic Entomology*. 50, no. 6 (1957): 778–784.

Saito, T., and S. Bjørnson. 2006. "Horizontal Transmission of a Microsporidium from the Convergent Lady Beetle, *Hippodamia convergens* Guérin-Méneville (Coleoptera: Coccinellidae), to Three Coccinellid Species of Nova Scotia." *Biological Control* 39, no. 3 (2006): 427–433.

Schaefer, Carl W., and Antônio R. Panizzi, eds. *Heteroptera of Economic Importance*. CRC Press, 2000.

Shrewsbury, Paula M., and Michael J. Raupp. "Do Top-down or Bottom-up Forces Determine *Stephanitis pyriodes* Abundance in Urban Landscapes?" *Ecological Applications* 16, no. 1 (2006): 262–272.

Stireman, John O. III, James E. O'Hara, and D. Monty Wood. "Tachinidae: Evolution, Behavior, and Ecology." *Annual Review of Entomology* 51 (2006): 525–555.

Tallamy, Douglas W. *Bringing Nature Home: How You Can Sustain Wildlife with Native Plants*, rev. ed. Timber Press, 2009.

Tillman, P. G., H. A. Smith, and J. M. Holland. "Cover Crops and Related Methods for Enhancing Agricultural Biodiversity and Conservation Biocontrol: Successful Case Studies." In *Biodiversity and Insects Pests: Key Issues for Sustainable Management*, edited by Geoff M. Gurr, Steve D. Wratten, William E. Snyder, and Donna M. Y. Read, 309–328. John Wiley, 2012.

Tooker, John F., and Lawrence M. Hanks. "Influence of Plant Community Structure on Natural Enemies of Pine Needle Scale (Homoptera: Diaspididae) in Urban Landscapes." *Environmental Entomology* 29, no. 6 (2000): 1305–1311.

Tscharntke, Teja, Riccardo Bommarco, Yann Clough, Thomas O. Crist, David Kleijn, Tatyana A. Rand, Jason M. Tylianakis, Saskya van Nouhuys, and Stefan Vidal. "Conservation Biological Control and Enemy Diversity on a Landscape Scale." *Biological Control* 43, no. 3 (2007): 294–309.

FOR MORE INFORMATION . . .

A great source for more information about designing and installing farm buffers is the USDA National Agroforestry Center (NAC). The NAC offers a variety of print publications, website tools, and free software for conservation engineers and farmers.

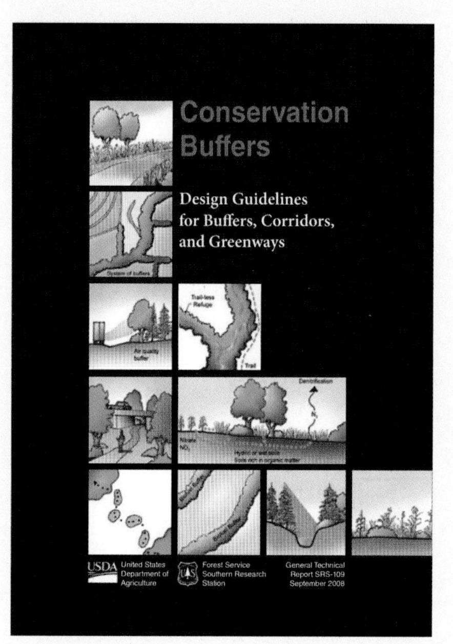

Among these resources is their groundbreaking book *Conservation Buffers: Design Guidelines for Buffers, Corridors, and Greenways*, authored by landscape planner Gary Bentrup. The book includes more than 80 illustrated designs for various types of buffers that protect biodiversity, reduce soil loss, enhance outdoor recreation, maintain water quality, and achieve other conservation goals. Specific tables are included to help calculate the optimal size of buffer features for various intended purposes. The book is available free from the NAC in print form, and can be downloaded from their website. In addition to English, the guide is also available in Spanish, French, Chinese, Korean, and even Mongolian!

Along with the *Conservation Buffers* guide, the NAC produces free software for conservation planners, including BUFFER$, an Excel-based cost-benefit analysis tool for calculating the economic trade-offs of establishing new buffers.

CanVis, another free software program, helps users create visual simulations of installed buffers. To use the software, you start with an actual photo of your farm or landscape, and add simulated landscape features, such as plants, to the photo from a library of stock images. The process helps stakeholders and landowners visualize what the established buffer could look like. The NAC even offers training videos on how to use the software.

To access these and other free resources from the NAC, visit their website at: http://nac.unl.edu.

About the Authors

Eric Lee-Mäder

Eric Lee-Mäder is Assistant Pollinator Program Director at the Xerces Society for Invertebrate Conservation and Extension Professor of Entomology at the University of Minnesota. In these roles Eric works with farmers and agencies like the USDA-NRCS to enhance functional biodiversity in working agricultural lands across the world — from his native North Dakota to India, where he wishes he were a native. Eric's professional background includes a mash-up of previous agroecology work including time spent as a small farm Extension educator, commercial beekeeper, and crop consultant for the native seed industry where he provided weekly insect and disease scouting on hundreds of plant species grown for prairie restoration efforts.

Jennifer Hopwood

Jennifer Hopwood, Pollinator Conservation Specialist at the Xerces Society for Invertebrate Conservation, provides resources and training to farmers and land managers for beneficial insect habitat management and restoration. Before joining Xerces, Jennifer was involved in ecological research and taught biology and environmental science. Her written work has appeared in scientific journals as well as popular publications, and she's presented to audiences across the United States. Jennifer earned her master's degree in entomology from the University of Kansas, and she is a Visiting Scholar with Michigan State University's Kellogg Biological Station.

Lora Morandin

Lora Morandin is an ecological consultant working out of Victoria, B.C., Canada. She completed a MSc at Western University, Canada, on pollination of greenhouse tomatoes by managed bumble bees, and a PhD at Simon Fraser University, Canada, on modern agriculture and pollination by native bees. Her latest work in agriculture was through the University of California, Berkeley where she studied effects of native plant hedgerows on beneficial and pest insects. She currently consults on varied ecological topics including agriculture and marine pollution. She's authored more than 15 peer-reviewed publications on native bees and natural enemies in agriculture.

Mace Vaughan

Mace Vaughan, Pollinator Program Director for the Xerces Society, earned his master's degrees in entomology and education from Cornell University. He has studied ground beetles in riparian forests of Utah and the behavior of honey bees in upstate New York, has wrangled insects for two PBS nature documentaries, and has taught a wide range of audiences across the United States about bees, spiders, and butterflies. Mace has led Xerces' Pollinator Conservation Program since 2003 and also serves as a joint Pollinator Conservation Specialist for the USDA Natural Resources Conservation Service. He is a co-author of the Storey book *Attracting Native Pollinators.*

Scott Hoffman Black

Scott Hoffman Black is Executive Director at the Xerces Society for Invertebrate Conservation in Portland, Oregon. He has extensive experience in endangered species conservation, pollinator conservation, and forest and range management issues. Scott has authored over 200 scientific and popular publications, coauthored two books and contributed chapters to many others. Scott has received several awards including the 2011 Colorado State University College of Agricultural Sciences Honor Alumnus Award and National Forest Service Wings Across Americas 2012 Butterfly Conservation Award.

The Xerces Society for Invertebrate Conservation

Protecting the Life That Sustains Us

Butterflies, dragonflies, beetles, worms, starfish, mussels, and crabs are but a few of the millions of invertebrates at the heart of a healthy environment. Invertebrates build the stunning coral reefs of our oceans; they are essential to the reproduction of most flowering plants, including many fruits, vegetables, and nuts; and they are food for birds, fish, and other animals. Yet invertebrate populations are often imperiled by human activities and rarely accounted for in mainstream conservation.

Established in 1971, the Xerces Society is at the forefront of invertebrate protection, harnessing the knowledge of scientists and public enthusiasm to implement conservation and education programs. Over the past three decades, we have protected endangered species and their habitats, produced groundbreaking publications on insect conservation, trained thousands of farmers and land managers to protect and manage habitat, and raised awareness about the invertebrates of forests, prairies, deserts, and oceans.

To continue this work, the Xerces Society needs your support. By joining Xerces, you can help give voice to these life-sustaining creatures.

OUR WORK

Of the more than one million species of animals in the world, 94 percent are invertebrates. The services they perform — pollination, seed dispersal, food for wildlife, nutrient recycling — are critical to life on our planet. Indeed, without them whole ecosystems would collapse. But when decisions are made about environmental policy and land management, these vital and diverse creatures are often overlooked. The Xerces Society works to address this situation in a variety of ways:

Education

We educate farmers, land managers, and the public about the importance of invertebrates by demonstrating that habitat protection and management are keys to their conservation. Our Pollinator Conservation Program trains farmers, agency officials, and park managers to protect, restore, and enhance areas for pollinators and other beneficial insects. Our Aquatic Conservation Program provides advice and resources to scientists, land managers, and watershed stewards for monitoring the health of streams, rivers, and wetlands.

Advocacy

Our Endangered Species Program advocates on behalf of threatened, endangered, and at-risk invertebrates and their habitats. From the world's rarest butterflies, to caddisflies that live only in one stream, to declining bumble bees, Xerces is dedicated to protecting invertebrates and the ecosystems that depend on them.

Policy

We work with federal agencies to incorporate the needs of pollinators and other invertebrates into national conservation programs. We work with lawmakers to pass legislation to improve habitat for invertebrates. We also promote invertebrate protection using the Endangered Species Act and other federal and state laws.

Xerces Society (continued)

Publications

Via our member magazine, *Wings: Essays on Invertebrate Conservation* — and through our books and website — we disseminate scientific information, updates on advocacy efforts, and practical suggestions for helping invertebrates. *Wings* has articles by leading conservationists and scientists and features extraordinary color images from renowned wildlife photographers. The society also publishes guidelines, fact sheets, and identification guides that help citizens take action to protect pollinators and other beneficial insects. Many of these publications are free on our website at: www.xerces.org.

SCIENTIFIC RESEARCH

We work on a variety of applied research projects that help us to protect habitat, ranging from how to effectively restore pollinator habitat on farms to understanding the life history of endangered species. Through the Joan Mosenthal DeWind Award, the Xerces Society offers grants to students conducting research on butterfly and moth conservation. Our staff members regularly write scientific papers and magazine articles, coordinate field work, and take leading roles in national and international scientific coalitions.

JOIN US

As a Xerces Society member, you will receive *Wings: Essays on Invertebrate Conservation* — our biannual color magazine — as well as timely bulletins on invertebrate conservation efforts. You will be eligible for discounts on books and merchandise. Most important, your support helps fund innovative conservation programs, effective education and advocacy, and scientific and popular publications, helping to spread the word about the vital role invertebrates play in our lives. Please join us today!

Index

Page numbers in *italics* indicate pictures and illustrations, those in **boldface** indicate tables, and a ***bold italic*** number means that an insect or plant profile appears on that page.

A

Achillea millefolium (yarrow), 54, 128, 133, 180, ***219***
Adamson, Nancy Lee, 170–171
Adjacent habitat, 51, *51*, 53
Ailanthus altissima (tree-of-heaven), 103
Alopecurus pratensis (meadow foxtail), 157
Alsike clover (*Trifolium hybridum*), 116, 128
Alternate pest host plants, 47, 48–49
Alternative prey, 48–49
Alyssum (*Lobularia maritima*), 54, 83, 87, 128, ***230***
Ambush (assassin) bugs, 5, **18–19**, *48*, ***184***
Amorpha fruticosa (false indigo), **222–223**
Anagrus spp. (fairyflies), 48
Anasa tristis (squash bug), 87
Andropogon gerardii (big bluestem), 145, ***226***, *227*
Anemone canadensis (Canada anemone), ***211***
Anethum graveolens (dill), 83, *83*, 87, ***231***
Angelica atropurpurea (purplestem angelica), ***217***
Annuals, insectary strips and, 84, **89–91**
Anthriscus cerefolium (chervil), 128
Aphidius colemani, 47
Aphids, 5, 17, 26, 47
Apocynum cannabinum (Indian hemp), ***214***
Appalachian Mountains, 172
Apple maggot (*Rhagoletis pomonella*), **55**
Apples, **55**, 166, 172–173
Arilus cristatus (wheel bugs), 184
Aristida spp. (wiregrass), ***228***, *229*
Arrowleaf clover (*Trifolium vesiculosum*), 181
Arthropods, predatory, 5, **18–19**
Asclepias spp., 28–29, *29*, 136, ***215***
Asian lady beetles (*Harmonia axyridis*), 17, *17*
Assassin bugs, 5, **18–19**, *48*, ***184***
Asters (*Symphyotrichum* spp.), ***210***
Augmentative biocontrol, 12–15

B

Baccharis spp. (coyotebrush), ***221***
Bachelor button (*Centaurea cyanus*), 83, ***230***
Bacillus thuringiensis (*Bt*), 162, 163–164
Backyard Habitat Program, 131
Bagrada bug, 87
Bagworms (*Thyridopteryx ephemeraeformis*), 7
Banker plants, 47
Baptisia spp. (wild indigo), 180
Barley, 47
Barriers, 166
Basswood (*Tilia americana*), ***220***
Beauty, 28
Beauveria bassiana, 164
Bed shapers, 146
Beebalm, lemon (*Monarda citriodora*), ***215***
Bees, 3, *24*, *163*
Beetle banks
- benefits of, 8, 50
- creating, 146–147, *146*, *147*, *148*
- effectiveness of, 145
- farm practices checklist and, 38
- Oregon case study, 156–157
- overview of, 144, *144*
- planning of, 145

Berms, *94*, 97–98, 146, *146*
Berries, **55**
Berseem clover (*Trifolium alexandrinum*), 116
Beyond the Birds and the Bees (Xerces), 162
Bicyrtes quadrifasciatus (sand wasps), *16*
Biddinger, David, 172
Big bluestem (*Andropogon gerardii*), 145, ***226***, *227*
Big-eyed bugs, **18–19**, ***185***
Biocontrol
- augmentative, 12–15

Biocontrol (*continued*)
beneficial insects for, 3–6
classic, 10–12
conservation, 15–17
farming and, 6–8
overview of, 1–2
Biocontrol agents, defined, 3
Blaauw, Brett, 128–129
Blackberry, Himalayan (*Rubus armeniacus*), 24, 51
Blackberry leafhoppers (*Dikrella cruentata*), 48
Black mustard (*Brassica nigra*), 120
Blanketflower (*Gaillardia* spp.), ***211***
Bluebells, California (*Phacelia campanularia*), *83*, 87
Blueberries, 128–129
Bluebunch wheatgrass (*Pseudoroegneria spicata*), 119, **226**, *227*
Bluestem. *See* Big bluestem; Little bluestem
Blue wild rye (*Elymus glaucus*), 119, 157, ***226***, *227*
Blue wild rye (*Festuca glauca*), 145
Boneset (*Eupatorium perfoliatum*), ***211***
Borders. *See also* Seeded borders
benefits of, *35*, 52
farm practices checklist and, 37
long-term maintenance, 66
overview of, 56–57
sample seed mixes for, 67
seeding methods, 61–66
site preparation, 58–60
Bouteloua curtipendula (sideoats grama), 119, ***228***, *229*
Boutelous dactyloides (buffalo grass), 125
Brassicas, as cover crops, 119–120, *119*
Broadcast seeders, 61, 64
Brush piles, 38, 153, *153*
Bt. See Bacillus thuringiensis
Buckwheat (*Fagopyrum esculentum*)
as cover crop, 120–121, *120*, 125, 127
in insectary strips, 83, 87
Kerr Center and, 181
overview of, ***231***
Buckwheat, wild (*Eriogonum* spp.), ***218***
Buffaloberry (*Shepherdia* spp.), ***220***
Buffalo grass (*Boutelous dactyloides*), 125
Buffers
benefits of, 24
contour, 132–133, *133*
Farm Practices Checklist and, 37
grassed waterways, 134
Iowa soybean aphid control case study, 141–142
organic farms and, 139–140
overview of, 130–131
pesticide drift and, 169
riparian, 135–136
shelterbelts and windbreaks, 137–138, *138*, *139*
Bumble bees, *163*
Burning, 66, *176*, 177–178, *177*
Buttonbush (*Cephalanthus occidentalis*), ***221***

C

Cabbage root maggot (*Delia radicum*), 120
Calendula (*Calendula* spp.), 83
California, **78–79**, **111**, 113–114
California bluebells (*Phacelia campanularia*), *83*, 87
California oatgrass (*Danthonica californica*), 145, ***226***, *227*
Callirhoe spp. (winecup), 181
Canada anemone (*Anemone canadensis*), ***211***
Canada wild rye (*Elymus canadensis*), 145, ***226***, *227*
Carabid beetle (*Scaphinotus marginatus*), 156
Carrot (*Daucus carota*), 128
Case studies
beetle banks in Oregon, 156–157
beneficial habitat on Oklahoma farm and ranch, 180–181
biological mite control in Pennsylvania apple orchard, 172–173
Christmas tree production, 40–41
cover crops on Michigan blueberry farm, 128–129
hedgerows on California farms, 113–114
milkweed, stink bugs, and Georgia cotton, 26–27
New Mexico pumpkins and insectary strips, 87–88
pest management in Washington vineyards, 26–27
windbreaks and pesticide drift, 170–171
Caterpillars, 45–46
Ceanothus spp. (wild lilac), ***224***
Cecropia moth (*Hyalophora cecropia*), *12*, 23
Centaurea cyanus, 83, ***230***
Cephalanthus occidentalis (buttonbush), ***221***
Chamaebatiaria millefolium (desert sweet), ***222***

Chamaecrista fasciculata, 54, *54*, 116, *116*, ***216***
Chauliognathus pennsylvanicus (goldenrod soldier beetles), *36*
Checkered beetles, ***191***
Chemical resistance, 8
Cherries, **55**
Cherry, wild (*Prunus* spp.), ***225***
Chervil (*Anthriscus cerefolium*), 128
Chicory (*Cichorium intybus*), 40
Chinavia hilaris (green stink bug), 28–29
Chionaspis pinifole (pine needle scale), 40
Christmas trees, 40–41
Chrysoperla spp. (green lacewing), 14, 190
Cichorium intybus (chicory), 40
Cilantro (*Coriandrum sativum*), 83, 128, ***231***
Classic biocontrol, overview of, 10–12
Clover, sweet (*Melilotus* spp.), 116, **233**
Clovers (*Trifolium* spp.)
  alsike, 116, 128
  arrowleaf, 181
  berseem, 116
  as cover crops, 116
  crimson, 116, *117*, 118, 125, *125*, 126, 128
  overview of, ***231***
  red, 116, *117*
  strawberry, 128
  white Dutch, 40, 125, 128, 131, 133
Coccinella septempunctata (seven-spotted lady beetle), 17
Coffeeberry (*Rhamnus californica*), ***221***
Common reed (*Phragmites australis*), 151
Compass plant (*Silphium perfoliatum*), *52*, 54, 151, ***212***
Complete metamorphosis, *4*, 183
Compsilura concinnata (tachinid flies), 12, *12*, **18–19**, 23, *182–183*, ***203***
Conservation biocontrol, 15–17. *See also* Habitat enhancements
Conservation buffers. *See* Buffers
Conservation Reserve Program (CRP), 177
Contour buffer strips, 132–133, *133*
Controlled grazing, 180
Convergent lady beetle (*Hippodamia convergens*), *3*, 13, *13*, 14, 88
Coppicing, 96
Coreopsis, *83*, 87, *91*, 134, 180, ***214–215***, ***217***
Coriander sativum (cilantro), 83, 128, ***231***
Corridors. *See* Movement corridors
Cosmopepla lintneriana, *48*
Cosmos bipinnatus, 87
Cotesia hyperparasitoid (*Hypopteromalus tabacum*), 46
Cotesia wasps, 46
Cotton, 7–8, 28–29
Cotton aphids (*Aphis gossypii*), 47
Cottony cushion scale (*Icerya purchasi*), 11, *11*, 17
Cover, *50*
Cover crop cocktails, *117*
Cover crops
  brassicas, 119–120, *119*
  buckwheat as, 120–121, *120*
  establishing, 122–123, *124*, 126, 127
  farm practices checklist and, 37
  grasses, 118–119, *118*, *119*
  lacy phacelia, 121–122, *121*
  legumes, 116, *116*, *117*
  overview of, 115–116
  small-scale, 125
  species descriptions, ***230–233***
  summer, **125**, 127
  Wind Dancer blueberry farm case study, 128–129
  winter, **125**, 126
Cowpea (*Vigna unguiculata*), 116, 181
Coyotebrush (*Baccharis* spp.), ***221***
Crataegus spp. (hawthorn), ***223***
Crimper-rollers, 123, *123*
Crimson clover (*Trifolium incarnatum*), 116, *117*, 118, 125, *125*, 126, 128
CRP. *See* Conservation Reserve Program
Cupplant (*Silphium perfoliatum*), *52*, 54, 151, ***212***

D

Daikon radish (*Raphanus sativus*), 119, *119*, **232**
Daisy fleabane (*Erigeron* spp.), 54, ***212***
Damsel bugs, **18–19**, ***186***
Danthonica californica (California oatgrass), 145, ***226***, *227*
Daucus carota. *See* Carrot; Queen Anne's lace
Deergrass (*Muhlenbergia rigens*), ***226***, *227*
Delia radicum (cabbage root maggot), 120
Desert sweet (*Chamaebatiaria millefolium*), ***222***
Diabrotica undecimpunctata howardi (spotted cucumber beetles), 87–88, *87*

Dikrella cruentata (blackberry leafhoppers), 48
Dill (*Anethum graveolens*), 83, *83*, 87, ***231***
Dipsacus fullonum (teasel), 151
Disking, 176–177
Diversity
food web complexity and, 46
habitat enhancements and, 47–48, 55
importance of, 40
of predators in field crop, 5–6
selecting for, 55
Dogbane (*Apocynum cannabinum*), ***214***
Dotted mint (*Monarda punctata*), ***212–213***
Downwash, 170
Drip irrigation, 99
Drop seeders, 61, 64
Drosophila suzukii (spotted-wing drosophila), 24, **55**
Duff layer, 156
Dutch clover. *See* White Dutch clover

E

Eastern gamagrass (*Tripsacum dactyloides*), *228*, ***228***
Economic thresholds, 10
Ecosystem services, 3, 141–142
Edge habitats, 52. *See also* Borders
Education, 28
Elderberry (*Sambucus* spp.), ***222***
Electrostatic sprayers, *168*, 169
Ellen, Gwendolyn, 156–157
Elymus canadensis (Canada wild rye), 145, ***226***, *227*
Elymus glaucus (blue wild rye), 119, 157, ***226***, *227*
Elymus trachycaulus (slender wheatgrass), 145, 157, ***228***, *229*
Environmental Quality Incentives Program (EQIP), 47
Erigeron spp. (daisy fleabane), 54, ***212***
Eriogonum spp. (wild buckwheat), ***218***
Erucic acid, 119
Erwinia amylovora (fire blight), **55**
Eryngium yuccifolium (rattlesnake master), 54, *136*, ***217***
Erythroneura elegantula (grape leafhoppers), 26, 48
Eupatorium perfoliatum (boneset), ***211***
European red mite, 172
Euthamia occidentalis (western goldentop), 136
Euthamia spp. *See* Goldenrod
Evergreens, 170–171, *171*
Exotic species. *See* Nonnative beneficial insects
Extrafloral nectar, 49

F

Fagopyrum esculentum (buckwheat). *See* Buckwheat
Fairyflies (*Anagrus* spp.), 48
False indigo (*Amorpha fruticosa*), ***222–223***
Farm practices checklist, 36–39
Fencerow revitalization, 102–103
Fennel (*Foeniculum vulgare*), 54
Fescue grass (*Festuca* spp.), 119, 145, ***228***, *229*
Field buffer systems. *See* Buffers
Filter strips, *134*, 136, *136*, *137*. *See also* Buffers
Fire, 66, *176*, 177–178, *177*
Fire blight (*Erwinia amylovora*), **55**
Firefly beetles, ***192***
Fish and Wildlife Structure Practice Standards, 150
Fleabane (*Erigeron* spp.), 54, ***212***
Flea beetles, 87
Flower flies. 7, **18–19**, 24, 45, *45*, *83*, ***199***
Flowers, 31, 36, 45–48, 53–54. *See also* Wildflowers
Foeniculum vulgare (fennel), 54
Food sprays, 49
Food supplements, 49
Food webs, 46
Forage radish (*Raphanus sativus*), 119, *119*, ***232***
Fruit bags, 166
Funding, 67

G

Gaillardia spp. (blanketflower), ***211***
Gamagrass, eastern (*Tripsacum dactyloides*), *228*, ***228***
Gardens. *See* Yards and gardens
Gathering Together Farm, 157
Generalists, 5
Georgia, 28–29
Giant silk moths (*Hyalophora cecropia*), *12*, 23
Globe gilia (*Gilia capitata*), ***213***
Glowworms, 192
Glucosinolates, 119
Golden Alexanders (*Zizia aurea*), *53*, 54, ***213***
Goldenrod (*Solidago* spp., *Oligoneuron* spp., *Euthamia* spp.), 54, *55*, ***213***
Goldenrod soldier beetles (*Chauliognathus pennsylvanicus*), *36*
Grape leafhoppers (*Erythroneura elegantula*), 26, 48

Grapes, **55**. *See also* Vineyards
Grass-carrying wasps (*Isodontia* spp.), *149*
Grassed waterways, 134
Grasses
beetle banks and, 144–145, *144*, 147, *147*, *148*, 157
big bluestem, 145, **226**, *227*
bluebunch wheatgrass, 119, **226**, *227*
blue wild rye, 145 , 157, **226**, *227*
for California, **78–79**
California oatgrass, 145, **226**, *227*
Canada wild rye, 145, **226**, *227*
contour buffer strips and, 132
as cover crops, 118–119, *118*, *119*
deergrass, **226**, *227*
eastern gamagrass, *228*, **228**
Idaho fescue, 119, **228**, *229*
Indian grass, *144*, 145, **228**, *229*
for intermountain west region, **76–77**
little bluestem, *118*, 119, *144*, **228**, *229*
for Midwestern U.S., **72–73**
for northeastern U.S., **68–69**
for northern plains region, **74–75**
overview of native, 226–229
Pacific northwest region, **80–81**
prairie dropseed, 145, **228**, *229*
prairie junegrass, *118*, 119, 145, **228**, *229*
Roemer's fescue, 145, **228**, *229*
sideoats grama, 119, **228**, *229*
slender wheatgrass, 145, 157, **228**, *229*
for southeastern U.S., **70–71**
for southern plains region, **76–77**
for southwestern U.S., **78–79**
wiregrass, **228**, *229*
Grass strips, *50*
Grasswitz, Tessa, 87–88
Grazing, 178, *178*, 180–181
Great ash sphinx moth caterpillar (*Sphinx chersis*), 46
Green lacewing (*Chrysoperla* spp.), 14, 190
Green manure crops, 116
Green stink bug (*Chinavia hilaris*), 28–29
Ground beetles, **18–19**, 156, ***193***
Gypsy moths (*Lymantria dispar*), 12, 23

H

Habitat
adjacent, 51, *51*
on farm, 35–36
flowers and, 36
overview of, 31–33
shelter and, 36
size and location of, 52, *53*
in surrounding landscape, 34
use of by lady beetles, 33
Habitat enhancements
alternate prey or hosts and, 48
diverse flower species and, 47–48
native plants and, 45–47
overview of, 8, 9, *9*
shelter and, 50
yards, gardens and, 7
Habitat management, overview of, 159
Hairy vetch (*Vicia villosa*), 116, 118, 125, 126, 181, **232**
Hand-broadcasting seed, 61, 62, *62–63*
Hand weeding, 66
Harlequin bug (*Murgantia histrionica*), 87
Harmonia axyridis (Asian lady beetles), 17, *17*
Harvestmen, **18–19**, *206*, ***206***
Hawaii, 23–24
Hawthorn (*Crataegus* spp.), **223**
Hedge-laying, 96
Hedgerows
adjacent habitat and, 51, *51*, *52*
costs of, 93
designing, 94–95, *94–95*
farm practices checklist and, 37
at home, 99
overview of, 92–93, *93*, 96
planning for, 96–99
planting of, 100–101
post-planting care, 101
property boundaries and, *32*
revitalizing old fencerows and, 102–103
sample plant mixes for, **104–112**
Heirloom varieties, 54
Helianthus annuus (sunflower), 181, ***218***, **233**
Hemaris diffinis (snowberry clearwing moth caterpillar), 46
Hemp, Indian (*Apocynum cannabinum*), ***214***
Herbicides, 60, 65
Heteromeles arbutifolia (toyon), ***224***
Heterorhabditis bacteriophora, 164–165

Himalayan blackberry (*Rubus armeniacus*), 24, 51
Hippodamia convergens. See Convergent lady beetle
Holodiscus discolor (ocean spray), **223**
Home gardens. *See* Yards and gardens
Honeybees, pesticides and, 162
Honeydew, 49, 54
Hornworm caterpillars (*Manduca* spp.), *6*, 46
Hosts, 6, 47, 48–49
Hotels, 38, 150, 154–155, *154*, *155*
Hover flies. *See* Flower flies
How to Reduce Bee Poisoning from Pesticides (Pacific Northwest Extension), 169
Hypepteromalus tabacum (cotesia hyperparasitoid), 46

I

Icerya purchasi (cottony cushiony scale), 11, *11*, 17
Idaho fescue (*Festuca idahoensis*), 119, **228**, *229*
Incomplete metamorphosis, *4*, 183
Indian blanket (*Gaillardia* spp.), ***211***
Indian grass (*Sorghastrum nutans*), *144*, 145, ***228***, *229*
Indian hemp (*Apocynum cannabinum*), ***214***
Indigo, false (*Amorpha fruticosa*), ***222–223***
Insectary strips
  New Mexico pumpkin case study, 87–88
  overview of, 82–83
  perennials vs. annuals for, 84
  planting of, 86
  sample seed mixes for, **89–91**
Insect hotels, 38, 150, 154–155, *154*, *155*
Insecticides. *See* Pesticides
Insidious flower bug (*Orius insidiosus*). *See* Minute pirate bugs
Integrated Pest Management (IPM), 10, 161, 167, 169
Intermountain west region, **76–77**, **109**
Interseeding, 66, *66*, 95, 179
Invasive species, 51, 102–103
Iowa, 141–142
IPM. *See* Integrated Pest Management
Ironweed (*Veronia* spp.), 180, ***214***
Irrigation, hedgerows and, 94–95, 99, 100
Isaacs, Rufus, 45, 128–129
Isodontia spp. (grass-carrying wasps), *149*

J

James, David G., 26–27
Japan, 166
Jumping spiders, **204–205**, *205*
Junegrass (*Koeleria macrantha*), *118*, 119, 145, ***228***, *229*

K

Kaolin clay, *166*, 181
Kerr Center for Sustainable Agriculture, 180–181, *181*
Koeleria macrantha (prairie junegrass), *118*, 119, 145, ***228***, *229*
Kremin, Claire, 93
Kuepper, George, 180–181

L

Lacewings. *See also* Green lacewing
  cover crops and, 128
  life cycle of, *4*
  New Mexico pumpkin case study and, 88
  overview of, **18–19**, ***190***
  quantity of aphids consumed, 5
Lacy phacelia (*Phacelia tanacetifolia*), 83, 121–122, *121*, ***232***
Lady beetles. *See also* Vedalia beetle
  Asian, 17, *17*
  as beneficials, 3, *3*
  convergent, 3, 13, *13*, 14, 88
  habitat use by, 33
  milkweed aphids and, *48*
  native vs. exotic, 17
  overview of, **18–19**, 195
  seven-spotted, 17
  Stethorus, 172
Lanceleaf coreopsis (*Coreopsis lanceolata*), 134, **214–215**
Landis, Doug, 24, 45
Landscape fabric, 100
Larval stages, *4*
Leaf-footed bug (*Leptoglossus phyllopus*), 29
Leafhoppers, blackberry (*Dikrella cruentata*), 48
Leafhoppers, grape (*Erythroneura elegantula*), 26, 48
Leafroller caterpillars, **55**
Legumes, 116, *116*, *117*, 132. *See also Specific legumes*
Lemon beebalm (*Monarda citriodora*), ***215***
Leptoglossus phyllopus. See Leaf-footed bug
Lerew Brothers Farm, 172–173
Lettuce, 82
Life cycles, *4*
Lightning bugs, ***192***

Lilac, wild (*Ceanothus* spp.), ***224***
Limnanthes spp. (meadowfoam), ***215***
Little bluestem (*Schizachyrium scoparium*), *118*, 119, *144*, **228**, *229*
Livestock, 178, *178*, 180–181
Lobularia maritima (alyssum), 54, 83, 87, 128, ***230***
Long, Rachel, 93
Long-horned bees (*Melissodes* spp.), *24*
Long-term management, overview of, 174–175
Lygus bugs, 7–8, 121
Lymantria dispar (gypsy moths), 12, 23

M

Mallow, 51
Managed fire, 66, *176*, 177–178, *177*
Management-intensive grazing, 180
Managing Cover Crops Profitably (USDA), 122–123
Mangora spp. (orb weaver spiders), 5, *204*
Mantids, 5, ***189***
Marigold (*Tagetes* spp.), 83
Meadowfoam (*Limnanthes* spp.), ***215***
Meadow foxtail (*Alopecurus pratensis*), 157
Mechanical drop seeders, 61, 64
Melissodes spp. (long-horned bees), *24*
Metamorphosis, *4*, 183
Mexican hat (*Ratibida columnifera*), 180–181
Michigan, 128–129
Microbial insecticides, 163–165, 172–173
Midwestern U.S., **72–73, 106**
Milkweed (*Asclepias* spp.), 28–29, *29*, 136, ***215***
Milkweed aphids, *48*
Mint, 212–213, ***216***
Minute pirate bugs, 3, *4*, 14, **18–19**, ***187***
Mites, 5, 172–173. *See also* Predatory mites
Mobility, habitat colonization and, 34
Monarda citriodora (lemon beebalm), 215
Monarda punctata (dotted mint), ***212–213***
Morandin, Lora, 93
Mountain mint (*Pycnanthemum* spp.), ***216***
Movement corridors, 37–38
Mowing, 40, 65–66, 177
Muhlenbergia rigens (deergrass), ***226***, *227*
Mulch, 95, 150
Murgantia histrionica (harlequin bug), 87
Mustards (*Brassica* spp. and *Sinapis* spp.), 51, 119, 120, ***232***

N

Nasturtium (*Tropaeolum majus*), 128
National Agroforestry Center (NAC), 140
National Organic Program (NOP), 25, 26, 139
National Organic Standards Board, 24
Natural enemies, defined, 3
Natural Resources Conservation Service (NRCS), 47, 57, 67, 131, 161, 167
Neal Smith Wildlife Refuge, 142
Nectar, 32, 49, 54, *54*, 82, 120–121
Neem, 181
Neighbors, 103
Nematodes, 119, 163–165
Neonicotinoid pesticides, 162–163
Nest blocks, 151
Nesting, 38, *50*, 100, 151–152. *See also* Tunnel nests
New Mexico, 87–88
Nezara viridula (southern green stink bug), 28–29
Nonnative beneficial insects. *See also Specific insects*
classical biocontrol and, 10–12
commonly used, 14
dangers of introducing, 17
reducing use of, 23–24
Nonnative plant species
beneficial, ***230–233***
habitat enhancements and, 23–24
in hedgerows, 94
pests and parasitoids in, *53*
revitalization of hedgerows and, 102–103
Northeastern U.S., **68–69, 104**
Northern plains region, **74–75, 107**
No-till seed drills, 61, 65, *65*, 123, 179
No-till systems, *31*
Nozzles, pesticide drift and, 168–169, 171
NRCS. *See* Natural Resources Conservation Service
Nuisance plants, avoiding, 46–47
Nymphal stages, *4*

O

Oatgrass, California (*Danthonica californica*), 145, ***226***, *227*
Ocean spray (*Holodiscus discolor*), ***223***
Oilseed radish (*Raphanus sativus*), 119, *119*, ***232***
Oklahoma, 180–181
Oligoneuron spp. *See* Goldenrod

O'Neal, Matt, 141–142
Opposite plowing, 146, *146*, *147*
Orb weaver spiders (*Mangora* spp.), *5*, *204*
Oregon, 156–157
Organic farms, 24–25, 26, 28, 82, 139–140, 163
Orius insidiosus. See Minute pirate bugs

P

Pacific northwest region, **80–81**, **112**
Panicum virgatum (switchgrass), 145
Paper wasps (*Polistes* spp.), 150
Parasitoid insects
  as beneficials, 3, *3*
  in native vs. nonnative plants, 53
  overview of, 6, *6*, **20–21**
  tachinid flies, 12, *12*, **18–19**, 23, *182–183*, ***203***
  wasps, 6, *6*, *7*, **18–19**, ***201***, ***202***
Parsley (*Petroselinum crispum*), 128
Partridge pea (*Chamaecrista fasciculata*), 54, *54*, 116, *116*, ***216***
Peaches, **55**
Peanuts, 28–29
Pears, **55**, 166
Pennsylvania, 172–173
Pennsylvania State University Fruit Research and Extension Center, 172–173
Perennials, insectary strips and, 84, **89–91**
Perovskia atriplicifolia (Russian sage), 94
Persephone Farm, 157
Pesticide drift
  controlling, 167–169
  hedgerows and, 96
  native plant borders and, 57
  windbreaks and, 138, 139–140, *139*, 170–171
Pesticides
  alternatives to, 166–167
  classical biocontrol as alternative to, 11
  contour buffer strips and, 133
  farming and, 6–9
  farm practices checklist and, 36–37
  insectary strips and, 82, 83
  microbial, 163–165
  neonicotinoids, 162
  organic, 181
  reducing impacts from, 160–161
  reducing use of, 9, 23
  resistance to, 172
  secondary pests and, 7–8
  selection of, 162–163, 169
  switching, 9
  use of, 161–162
Phacelia spp., 83, *83*, 87, 121–122, *121*, ***216***, ***232***
Pheromone traps, *166*, 167
Phragmites australis (common reed), 151
Phytoseiulus persimilis (a predatory mite), 14
Pine needle scale (*Chionaspis pinifole*), 40
Pirate bugs, 88
Plains coreopsis (*Coreopsis tinctoria*), 83, 87, *91*, ***217***
Plant mixes. *See also* Seed mixes
  for California, **111**
  California farms case study, 113–114
  for hedgerows, **104–112**
  for intermountain west region, **109**
  for Midwestern U.S., **106**
  for northeastern U.S., **104**
  for northern plains region, **107**
  for Pacific northwest, **112**
  for southeastern U.S., **105**
  for southern plains region, **108**
  for southwestern U.S., **110**
Plum, wild. *See* Wild plum
Polistes spp. (paper wasps), 150
Pollen, 32, 49, 82
Pollinator Conservation Program, 131
Pollinators, 3, 24
Porosity, 170
Practice standards, 145
Prairie dropseed (*Sporobolus heterolepsis*), 145, ***228***, *229*
Prairie habitats, 141–142
Prairie junegrass (*Koeleria macrantha*), *118*, 119, 145, ***228***, *229*
Praying mantids, 5, ***189***
Predatory insects. *See also* Lady beetles
  assassin bugs/ambush bugs, *5*, **18–19**, *48*, ***184***
  as beneficials, 3
  big-eyed bugs, **18–19**, ***185***
  checkered beetles, ***191***
  damsel bugs, **18–19**, ***186***
  firefly beetles, ***192***
  flower flies/hover flies/syrphid flies, **18–19**, 24, 45, *45*, *83*, ***199***
  ground beetles, **18–19**, 156, ***193***
  lacewings (green and brown), *4*, 5, 14, **18–19**, 88, 128, ***190***
  mantids, 5, ***189***
  minute pirate bugs/insidious flower bugs, 3, *4*, 14, **18–19**, ***187***
  overview of, 5–6
  rove beetles, **18–19**, ***198***

soft-winged flower beetles, ***196***
soldier beetles, **18–19**, *36*, *52*, 165, ***197***
stink bugs, *8*, 26–27, 28–29, *48*, ***188***
tiger beetles, ***194***
wasps, ***200***
Predatory mites, 14, **18–19**, 172–173, *173*, ***207***
Predatory wasps, **18–19**, ***200***
Prescribed burning, 66, *176*, 177–178, *177*
Prune (*Prunus domestica*), 48
Prunella vulgaris (selfheal), 133, 134, *135*, ***218***
Pruning, 100
Prunus spp., **225**
Pseudoroegneria spicata (bluebunch wheatgrass), 119, ***226***, *227*
Pterostichus spp., 156
Pumpkins, 87–88
Purplestem angelica (*Angelica atropurpurea*), ***217***
Pycnanthemum spp. (mountain mint), ***216***
Pyrethrin, 163, 181
Queen Anne's lace (*Daucus carota*), 40

R

Radish, forage (*Raphanus sativus*), 119, *119*, ***232***
Rain gardens, 131
Rangeland, 180–181
Rant, Richard, 128–129
Rapeseed (*Brassica napus*), 119
Ratibida columnifera (Mexican hat), 180–181
Rattlesnake master (*Eryngium yuccifolium*), 54, *136*, ***217***
Recreation, 28
Red clover (*Trifolium pratense*), 116, *117*
Redhage, David, 180–181
Reduced tillage systems, *31*
Reed, common (*Phragmites australis*), 151
Remnant habitat patches, *34*, *159*
Residues, cover crops and, 123
Reverse plowing, 146, *146*, *147*
Rhagoletis pomonella (apple maggot), **55**
Rhamnus californica (coffeeberry), ***221***
Rhopalosiphum padii, 47
Riparian buffers, 135–136, *137*
Rodolia cardinalis (vedalia beetle), 11, *11*, 17
Roemer's fescue (*Festuca idahoensis roemerii*), 145, ***228***, *229*
Rosa spp., 26, **225**
Rosemary (*Rosmarinus officinalis*), 94
Roses, 26, **225**
Rosinweed (*Silphium perfoliatum*), *52*, 54, 151, ***212***
Rotating habitat disturbance, 179
Rotational grazing, 180
Rove beetles, overview of, **18–19**, ***198***
Row covers, *166*
Rubus armeniacus (Himalayan blackberry), 24, 51
Russian sage (*Perovskia atriplicifolia*), 94
Rye (*Sercale cereale*), 181
Rye, blue wild (*Elymus glaucus*), 119, 157, ***226***, *227*
Rye, Canada wild (*Elymus canadensis*), 145, ***226***, *227*

S

Saccharopolyspora spinosa, 164
Sage, Russian (*Perovskia atriplicifolia*), 94
Salix spp. (willow), **225**
Sambucus spp. (elderberry), **222**
Sand wasps (*Bicyrtes quadrifasciatus*), *16*
SARE. *See* Sustainable Agriculture Research and Education
Scale, 11, *11*, 17, 40
Scaphinotus marginatus (carabid beetle), 156
Scarab-hunting wasps, ***202***
Schizachyrium scoparium (little bluestem), *118*, 119, *144*, ***228***, *229*
Secondary pests, insecticides and, 7–8
Seed drills, 61, 65, *65*, 123, 179
Seeded borders. *See* Borders
Seed mixes. *See also* Plant mixes
for California, **78–79**
for eastern U.S., **89**
for insectary strips, **89–91**
for intermountain west region, **76–77**
for Midwestern U.S., **72–73**
for native plant borders, **68–81**
for northeastern U.S., **68–69**
for northern plains region, **74–75**
for Pacific northwest region, **80–81**
for southeastern U.S., **70–71**
for southern plains region, **76–77**
for southwestern U.S., **78–79**
for western U.S., **90–91**
Selfheal (*Prunella vulgaris*), 133, 134, *135*, ***218***
Sercale cereale (rye), 181

Seven-spotted lady beetle (*Coccinella septempunctata*), 17
Sheet-weaving spiders (*Linyphiidae*), 48, ***204–205***
Shelter. *See also* Beetle banks
  brush piles, 153, *153*
  habitat and, 36
  habitat enhancements and, 50, *50*
  importance of, 31
  insect hotels, 154–155, *154*, *155*
  lady beetles and, *33*
  Oregon beetle banks case study, 156–157
  overview of, 143–144
  riparian buffers and, 136
  tunnel nests, 149, *149*
  for wasps, 150–152
Shelterbelts, 137–138, *139*
Shepherdia spp. (buffaloberry), ***220***
Shrubs, native, ***220–225***
Sideoats grama (*Bouteloua curtipendula*), 119, ***228***, *229*
Silk moth, giant (*Hyalophora cecropia*), *12*, 23
Silphium perfoliatum (cup-plant), *52*, 54, 151, ***212***
Site preparation, 58–60
Slender wheatgrass (*Elymus trachycaulus*), 145, 157, ***228***, *229*
Smother crops, 120–121
Snowberry clearwing moth caterpillar (*Hemaris diffinis*), 46
Soft-winged flower beetles, ***196***
Soil solarization, *58*, 60, *60*, 99
Solarization, *58*, 60, *60*, 99
Soldier beetles, **18–19**, *36*, *52*, 165, ***197***
Solidago spp. *See* Goldenrod
Sorghastrum nutans (Indian grass), *144*, 145, ***228***, *229*
Sorghum x drummondii (sudangrass), 181
Southeastern U.S., **70–71**, **105**
Southern green stink bug (*Nezara viridula*), 28–29
Southern plains region, **76–77**, **108**
Southwestern U.S., **78–79**, **110**
Soybean aphids, 24, 141–142
Sphinx chersis (great ash sphinx moth caterpillar), 46
Spider mites, 27
Spiders
  as beneficials, 3, 5, 204–205
  jumping, ***204–205***, *205*
  orb weaver, 5, *204*
  overview of, **18–19**
  sheet-weaving, 48, ***204–205***
  wolf, 46, ***204–205***, *205*
Spinosad, 163, 164
Spirea spp., ***224***
Sporobolus heterolepsis (prairie dropseed), 145, ***228***, *229*
Spot spraying, 65
Spotted cucumber beetles (*Diabrotica undecimpunctata howardi*), 87–88, *87*
Spotted-wing drosophila (*Drosophila suzukii*), 24, **55**
Spray drift. *See* Pesticide drift
Springtails, 48
Squash bug (*Anasa tristis*), 87
Steinerma spp., 165
Stem bundles, 151–152, *151*
Stethorus lady beetles, 172
Stink bugs, *8*, 26–27, 28–29, *48*, ***188***
Strawberry clover (*Trifolium fragiferum*), 128
String trimming, weed control and, 65
STRIPS project, 141–142
Sudangrass (*Sorghum x drummondii*), 181
Sulfur sprays, 27
Sunflower (*Helianthus annuus*), 181, ***218***, ***233***
"Super" pests, 8
Sustainable Agriculture Research and Education (SARE), 123
Swamp milkweed (*Asclepias incarnata*), 136
Sweet clover (*Melilotus* spp.), 116, **233**
Switchgrass (*Panicum virgatum*), 145

T

Tachinid flies (*Compsilura concinnata*), 12, *12*, **18–19**, 23, ***203***
Tagetes spp. (marigold), 83
Tallamy, Doug, 45
Teasel (*Dipsacus fullonum*), 151
Temperature, pesticide drift and, 168
Termination, cover crops and, 123, *123*
Thyridopteryx ephemeraeformis. *See* Bagworms
Tiger beetles, ***194***
Tilia americana (basswood), ***220***
Tillage systems, *31*, 32, *32*, 37, 58
Tillman, Glynn, 28–29
Tobacco, 150
Tomatoes, *51*, 93, *93*
Tooker, John F., 40–41
Tower sprayers, *168*
Toyon (*Heteromeles arbutifolia*), ***224***
Tree-of-heaven (*Ailanthus altissima*), 103

Trees, native, ***220–225***
Trichogramma brassicae, 14
Trichopoda pennipes, 29
Trifolium spp. *See* Clovers
Tripsacum dactyloides (eastern gamagrass), *228*, ***228***
Tropaeolum majus (nasturtium), 128
True bugs, life cycle of, *4*
Tunnel nests, 149, *149*
Turf grasses, bagworms and, 7
Typhlodromus pyri, 172–173

U

University of California Integrated Pest Management Program, 169
Upland Wildlife Habitat Management Program, 150
USDA Natural Resources Conservation Service (NRCS), 47, 57, 67, 131, 161, 167

V

Vaughan, Mace, 99, 170–171
Vedalia beetle (*Rodolia cardinalis*), 11, *11*, 17
Verbesina spp. (wingstem), ***219***
Veronia spp. (ironweed), 180, ***214***
Vetch (*Vicia* spp.), 116, 118, 125, 126, 181, **232**
Vigna unguiculata (cowpea), 116, 181
Vineyards, 17, 26–27, 48
Viruses, insect, 164

W

Ward, Thomas, 170–171
Washington, vineyards in, 26–27
Wasps
- *Aphidius colemani*, 47
- as beneficials, 3, *3*
- nectar-collecting, 54
- nonnative, 14
- parasitoid, 6, *6*, *7*, **18–19**, ***201***, ***202***
- predatory, **18–19**, ***200***
- sand, *16*
- shelter for, 149, 150–152
- social, 151
- solitary, 151–152
- tunnel nests and, 149

Waterways, grassed, 134
Weather, pesticide drift and, 168, 171
Weeds
- adjacent habitat and, 51
- buckwheat and, 120–121
- control of during seedling establishment, 64–66
- hedgerows and, 98, 100
- lacy phacelia and, 121
- long-term management and, 175
- neonicotinoid pesticides and, 162
- site preparation and, 57–60

Western goldentop (*Euthamia occidentalis*), 136
Wheat, 47
Wheatgrass, 119, 145, 157, ***226***, *227*, ***228***, *229*
Wheel bugs (*Arilus cristatus*), 184
White Dutch clover (*Trifolium repens*), 40, 125, 128, 131, 133
White malady (*Chionaspis pinifole*), 40
Wild buckwheat (*Eriogonum* spp.), ***218***
Wild cherry (*Prunus* spp.), ***225***
Wildflowers. *See also* Flowers
- asters, ***210***
- bagworms and, 7
- beetle banks and, 144
- blanketflower/Indian blanket, ***211***
- boneset, ***211***
- for California, **78–79**
- Canada anemone, ***211***
- cupplant/compass plant/rosinweed, *52*, 54, 151, ***212***
- daisy fleabane, 54, ***212***
- dotted mint, ***212–213***
- globe gilia, ***213***
- golden Alexanders, *53*, 54, ***213***
- goldenrod, 54, *55*, ***213***
- Indian hemp/dogbane, ***214***
- for intermountain west region, **76–77**
- interseeding of, 179
- ironweed, 180, ***214***
- lanceleaf coreopsis, 134, ***214–215***
- lemon beebalm, ***215***
- meadowfoam, ***215***
- for Midwestern U.S., **72–73**
- milkweed, 26–27, 28–29, *29*, 136, ***215***
- mountain mint, ***216***
- for northeastern U.S., **68–69**
- for northern plains region, **74–75**
- overview of native, 210
- for Pacific northwest region, **80–81**
- partridge pea, 54, *54*, 116, *116*, ***216***
- phacelia, 83, *83*, 87, 121–122, *121*, ***216***, ***232***
- plains coreopsis, *83*, 87, *91*, ***217***
- purplestem angelica, ***217***
- rattlesnake master, 54, *136*, ***217***
- selection of, 53–54
- self-heal, 133, 134, *135*, ***218***
- for southeastern U.S., **70–71**
- for southern plains region, **76–77**

Wilflowers (*continued*)
for southwestern U.S., ***78–79***
sunflower, 181, ***218***, ***233***
wild buckwheat, ***218***
wingstem, ***219***
yarrow, 54, 128, 133, 180, ***219***
Wild indigo (*Baptisia* spp.), 180
Wild lilac (*Ceanothus* spp.), ***224***
Wild mustard, 51
Wild plum (*Prunus* spp.), ***225***
Wild rose (*Rosa* spp.), 26, ***225***
Wild rye. *See* Blue wild rye; Canada wild rye
Willow (*Salix* spp.), ***225***
Wind, pesticide drift and, 168, 171
Windbreaks, 137–138, *138*, 169, 170–171
Wind Dancer blueberry farm, 128–129
Winecup (*Callirhoe* spp.), 181
Wingstem (*Verbesina* spp.), ***219***
Winter cover, 50
Wiregrass (*Aristida* spp.), ***228***, 229
Wolf spiders, 46, ***204–205***, *205*

Y

Yards and gardens
beneficial insects in, 24
considerations for, 35, 54–55
cover crops and, 125
habitat enhancements and, 7
hedgerows and, 99
insectary plants in, 87
native borders for, 67
pesticides and, 166, 167
rain gardens and, 131
shelters in, 150
Yarrow (*Achillea millefolium*), 54, 128, 133, 180, ***219***

Z

Zelus spp., 5
Zizia aurea (golden Alexanders), 53, 54, ***213***

Photo Credits

Achim Raschka, Wikimedia Commons: 52 (bottom)
© Acterra: Action for a Healthy Planet: 153
© alacatr/iStockphoto.com: 161
Alex W. Covington, Wikimedia Commons: 225 (top right)
© Alan Shadow, NRCS: 70 (center), 116, 225 (top left)
© Alex Wild: x (top), 3 (top), 6 (top), 48 (bottom), 182, 191 (top), 195 (top)
© Alexandre Dulaunoy: 154
© Allen Casey, NRCS: 69 (center), 216 (bottom)
© Amanda Morse: 166 (bottom right)
© Andrew Dunham, Grinnell Heritage Farm: 143, 234
© Anna MacDonald: 141
© Banks Photos/iStockphoto.com: 25 (left)
© Betsy Betros: 3 (bottom), 184 (top), 195 (bottom), 202 (right)
© Bill Bouton: 203 (left)
© Bonnie L. Harper, Lady Bird Johnson Wildflower Center: 229 (middle left)
Bradley Higbee, Paramount Farming, Bugwood.org: 23, 185 (bottom)
© Brianna Borders, The Xerces Society: x (bottom), xi, 59, 62 (left), 63 (top left & right, bottom left)
© Bruce MacQueen/iStockphoto.com: xiv
Bruce Marlin, Wikimedia Commons: 198 (middle)
© Bruce Newhouse: 77 (middle), 79 (top), 80 (right), 118 (top), 213 (top), 227 (middle right)
© Bryan E. Reynolds: 36, 190, 193, 194 (all), 202 (left), 205 (both), 210, 211 (top)
© Carl Kurtz: 199 (left)
© Cathy Keifer/iStockphoto.com: 12
Charles J. Sharp, Wikimedia Commons: 206 (bottom)
Charles T. Bryson, USDA Agricultural Research Service, Bugwood.org: 217 (top)
Chhe, Wikimedia Commons: 72 (right), 118 (middle)
Chris Evans, Illinois Wildlife Action Plan, Bugwood.org: 221 (top right), 222 (bottom right)
© Chris Helzer: 34 (bottom), 88, 174, 177
© Cliff1066, flickr.com: 232 (bottom right)
© Colette Kessler, USDA-SD NRCS: 32 (top)
Crazytwoknobs, Wikimedia Commons: 70 (right), 217 (middle), 227 (bottom left)
© Dave McShaffrey: 16
Dave Powell, USDA Forest Service, Bugwood.org: 72 (center), 229 (middle right & bottom center)
David Cappaert, Michigan State University, Bugwood.org: 5 (top), 7 (top), 53, 192, 206 (top)
© David J. Biddinger: 173, 207 (bottom)
© Dr. David G. James: 13, 187, 201
© Dean Pennala Photography/iStockphoto.com: 42
© Debbie Roos: ii, 6 (bottom three), 8, 184 (bottom), 188 (bottom)
© J. Derek Scasta: 178
© Don Keirstead, NH NRCS: ix, 22, 56, 57
© Dot Paul, NRCS: 171
© Dr. Blake Brown: 31
Eric Hunt, Wikimedia Commons: 80 (center), 215 (bottom)
© Eric Lee-Mader, The Xerces Society: 26, 62 (right), 68 (center), 69 (left), 77 (bottom), 81 (left), 101, 117 (top), 118 (bottom), 119 (top), 120, 123 (bottom), 125, 213 (middle & bottom), 229 (top center)

© Ernst Conservation Seeds: 69 (right)
Evangele19, Wikimedia Commons: 221 (bottom)
Forest & Kim Starr, Starr Environmental, Bugwood.org: 230 (top)
Gabriel Hurley, Wikimedia Commons: 221 (top)
© Gary A. Monroe: 229 (top left)
© Gene Barrickman: 149 (top)
Gerald J. Lenhard, Louisiana State University, Bugwood.org: 5 (bottom)
© Gwendolyn Ellen: 144 (top & bottom), 148 (all), 157
© Heather Holm: 149 (bottom)
Howard F. Schwartz, Colorado State University, Bugwood.org: 75 (left), 227 (top right), 229 (top right & bottom left), 231 (bottom left & right), 233 (left)
Huw Williams, Wikimedia Commons: 55
James H. Miller, USDA Forest Service, Bugwood.org: 54, 214 (top left), 222 (bottom left), 223 (left)
Jamie Nielsen, University of Alaska Fairbanks, Cooperative Extension Service, Bugwood.org: 233 (right)
© Jeff Vanuga, NRCS: 102, 164
Jeffrey W. Lotz, Florida Department of Agriculture and Consumer Services, Bugwood.org: 11
© Jennifer Anderson @ USDA-NRCS PLANTS Database: 227 (top left)
© Jennifer Hopwood, The Xerces Society: xiii, 34 (top), 44, 50 (top), 68 (left), 71 (left), 75 (center), 90, 91, 136 (top), 158, 180, 181, 196 (top), 212 (bottom)
© Jessa Cruz, The Xerces Society: vi, 15, 32 (bottom), 49, 52 (top), 63 (bottom right), 65, 66 (top & bottom), 82, 84, 85, 92, 97, 98, 100, 176 (bottom)
© Jill Sidebottom: 41
© Jill Welham, Mirrored Images: 155
© John Anderson, Hedgerow Farms, Inc.: 29, 35, 74 (right), 79 (2nd from top), 130, 135 (bottom), 216 (middle), 232 (top right)
John D. Byrd, Mississippi State University, Bugwood.org: 135 (top)
© John Hartman: 169
John Ruberson, University of Georgia, Bugwood.org: 3 (middle)
John Ruter, University of Georgia, Bugwood.org: 224 (bottom)
Joseph Berger, Bugwood.org: 186 (top)
Joy Viola, Northeastern University, Bugwood.org: 220 (right)
Karan A. Rawlins, University of Georgia, Bugwood.org: 71 (right), 75 (right), 228
© Kathryn E. Bolin, Lady Bird Johnson Wildflower Center: 73 (left)
© Kevin Black, GROWMARK, Inc.: 7 (bottom), 24, 197 (bottom), 199 (center)
© Kimberly Gallagher: 179
© Kimiora Ward: 175
© Laura Westwood, The Xerces Society: 144 (middle)
© Lee Page, Lady Bird Johnson Wildflower Center: 214 (top right)
© Lorraine Seymour: 9, 27
Louis Tedders, USDA-ARS, Bugwood.org: 17
© Mace Vaughan, The Xerces Society: 25 (right), 80 (left), 134 (bottom), 136 (bottom)
Marvin Smith, Wikimedia Commons: 191 (bottom)
Mary Ellen (Mel) Harte, Bugwood.org: 74 (left), 81 (right)
© Matt Lavin, Wikimedia Commons: 81 (center), 222 (top)
© Matthew Shepherd, The Xerces Society: 30, 86, 121, 218 (top right), 223 (right), 225 (bottom)
Courtesy Michigan State University/Todd Martin: 115, 117 (middle & bottom), 119, 232 (top left & bottom left)
© Miguel Vieira, flickr.com: 224 (top right)
© MJ Hatfield: 79 (bottom), 188 (top), 197 (top), 199 (right), 212 (top left)
© Nancy Lee Adamson, The Xerces Society: 45, 48 (top), 58, 219 (left)
New Hampshire USDA-NRCS: 64
© Nikki Siebert, Low Country Local First: 71 (center)
© Norman G. Flaigg, Lady Bird Johnson Wildflower Center: 77 (top), 215 (top)
O. Pichard, Wikimedia Commons: 219 (right)
© Ori Chafe: 227 (middle left)
© Parker Deen/iStockphoto.com: 208
Pennsylvania Department of Conservation and Natural Resources – Forestry Archive: 198 (top)
© Peter Veilleux: 70 (left), 229 (bottom right)
© R. W. Smith, Lady Bird Johnson Wildflower Center: 217 (bottom)
© Rachel Long: 113
Rebekah D. Wallace, Colorado State University, Bugwood.org: 189
Rebekah D. Wallace, University of Georgia, Bugwood.org: 230 (bottom)
Richard Bartz, Munich Makro Freak, Wikimedia Commons: 186 (bottom)
© Rich Hatfield, The Xerces Society: 163
Richard Floyd, Creative Ideas LLC, Bugwood.org: 79 (3rd from top)
Rob Routledge, Sault College, Bugwood.org: 73 (right), 211 (middle & bottom), 214 (bottom)
© Rod Gilbert: 218 (top left & bottom), 224 (top left)
© Rollin Coville: 200
© Rufus Isaacs: 129
© Scott Seigfried: 68 (right), 72 (left), 73 (center), 74 (center), 204, 215 (middle), 216 (top)
Siga, Wikimedia Commons: 196 (bottom)
Sihem Cogneaux, Wikimedia Commons: 50 (bottom)
Stan Shebs, Wikimedia Commons: 227 (bottom right)
© Stephen P. Luk: 207 (top)
© Susan A. Beebe: i
© Tess Grasswitz: 83, 87, 151, 166 (bottom left), 231 (top left)
Thegreenj, Wikimedia Commons: 212 (top right)
© Toby Alexander, NRCS: 220 (left)
Tom Heutte, USDA Forest Service, Bugwood.org: 132 (bottom), 231 (top right)
© Tom Murray: 32 (middle), 198 (bottom)
USDA-ARS/Dale Spurgeon: 166 (top)
USDA-ARS/Jack Dykinga: 185 (top)
USDA-ARS/Keith Weller: 160, 168
USDA-ARS/Scott Bauer: 122, 191 (middle)
USDA-ARS/Stephen Ausmus: 10
USDA-ARS/Steven Mirsky: 123 (top)
USDA-NRCS: 241
USDA-NRCS/Erwin Cole: 140
USDA-NRCS/Lynn Betts: 131, 132 (top), 138
USDA-NRCS/Tim McCabe: 134 (top)
© Vilicus Farms/Anna Jones-Crabtree: 176 (top)
Whitney Cranshaw, Colorado State University, Bugwood.org: 2, 189 (bottom)
William M. Ciesla, Forest Health Management International, Bugwood.org: 203 (right)

ALSO BY THE XERCES SOCIETY

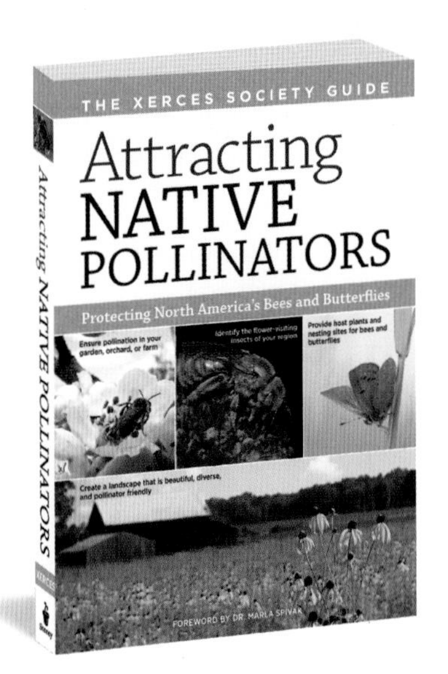

"This impressive reference should prove invaluable to gardeners, orchardists, farmers, park managers, and others responsible for the maintenance of green spaces."

— *Book News*

"Providing a thorough introduction to the interactions of plants and their essential insect partners, *Attracting Native Pollinators* . . . serves as an important reminder of how much we depend upon the natural world for our own survival."

— *The American Gardener*

384 pages.
Paper. ISBN 978-1-60342-695-4.

Other Storey Titles You Will Enjoy

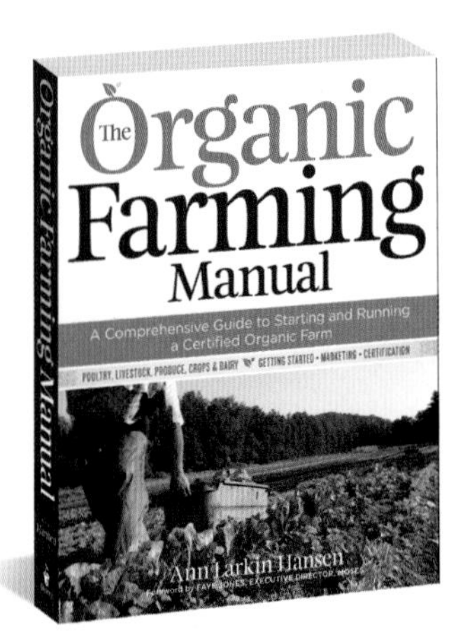

448 pages.
Paper. ISBN 978-1-60342-479-0.

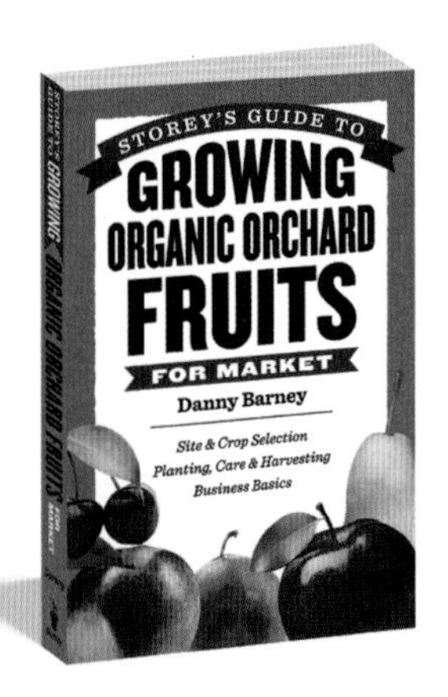

544 pages.
Paper. ISBN 978-1-60342-570-4.
Hardcover. 978-1-60342-723-4.

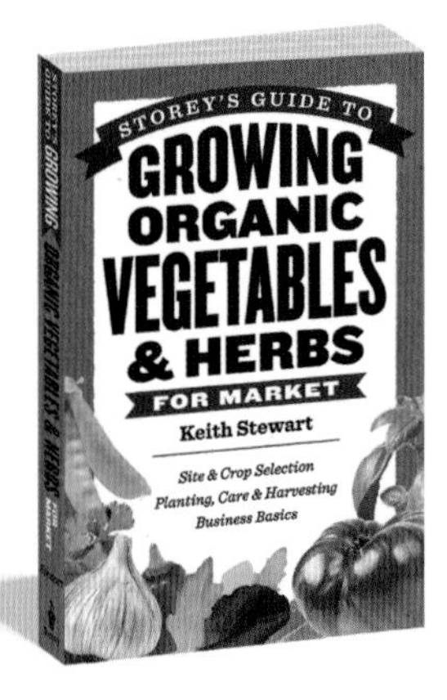

560 pages.
Paper. ISBN 978-1-60342-571-1.
Hardcover. 978-1-61212-007-2.

These and other books from Storey Publishing are available wherever quality books are sold or by calling 1-800-441-5700. Visit us at *www.storey.com* or sign up for our newsletter at *www.storey.com/signup*.